FRANKENSTEIN AND STEAM

FRANKENSTEIN AND STEAM

ESSAYS FOR CHARLES E. ROBINSON

EDITED BY ROBIN HAMMERMAN

UNIVERSITY OF DELAWARE | UNIVERSITY OF DELAWARE PRESS

NEWARK

Library of Congress Cataloging-in-Publication Data
Names: Hammerman, Robin, editor. | Robinson, Charles E., honouree.
Title: Frankenstein and STEAM: essays for Charles E. Robinson / edited by Robin Hammerman.
Description: Newark, DE: University of Delaware Press, [2022] | Includes bibliographical references and index.
Identifiers: LCCN 2021022262 | ISBN 9781644532522 (paperback) | ISBN 9781644532539 (hardback) | ISBN 9781644532546 (epub) | ISBN 9781644532553 (pdf)
Subjects: LCSH: Shelley, Mary Wollstonecraft, 1797–1851. Frankenstein. | Frankenstein's Monster (Fictitious character) | Literature and science. | Literature and technology. | LCGFT: Literary criticism.
Classification: LCC PR5397.F73 F688 2022 | DDC 823/.7—dc23
LC record available at https://lccn.loc.gov/2021022262

A British Cataloging-in-Publication record for this book is available from the British Library.

♾ The paper used in this publication meets the requirements of the American National Standard for Information Sciences—Permanence of Paper for Printed Library Materials, ANSI Z39.48-1992.

udpress.udel.edu

Distributed worldwide by Rutgers University Press

Manufactured in the United States of America

Contents

Introduction

Robin Hammerman

This collection of essays considers the vital importance of how an individual scholar may inspire current and future scholars to study and teach *Frankenstein* anew. Charles E. Robinson (1941–2016) was a teacher of distinction whose enthusiasm in the classroom was contagious. Charlie, as he liked to be called, was a generous mentor and a dynamic colleague. Anyone who studies *Frankenstein* knows of his contributions. His untimely passing, only two years before the novel's bicentennial in 2018, was a significant loss to the scholarly community. His scholarship, especially though not exclusively on *Frankenstein*, reveals admirable critical acumen, intuitive aplomb, and reverence for the literature that shaped the Romantic movement. Indeed, Charlie was one of the great Romanticists of our time.

This memorial volume is a partial testament to Charlie's enormous legacy. The essays, taken together, remind us to continue asking what it means to be human, a question that Mary Shelley's novel undoubtedly raised from its first publication in 1818 and that Charlie inspired generations of students and colleagues to explore. We aim to spotlight that question's significance here, at a time when developments in technoscience and new media challenge and blur boundaries between the animate and inanimate, when political leaders seek to dehumanize the most vulnerable among us, and when the Earth faces apocalyptic climate changes that may result in radical physical alterations to our species.

The two hundredth anniversary of Mary Shelley's novel *Frankenstein* in 2018 yielded a surge of interest in what was already one of the most widely read and translated novels written in English. The days leading to October 31, 2018, collectively branded "Frankenweek" by scholars and enthusiasts, were filled with live marathon "Frankenreads" of the novel, including the signature event hosted by the Library of Congress on Halloween.[1] Countless venues around the world hosted celebrations of the novel's legacy, including hundreds of college campuses, run by departments across disciplines spanning the arts and sciences. Events leading to the bicentennial throughout the year included conferences, symposia, art exhibitions, film screenings, and demonstrations of scientific and technological achievements in areas such as artificial intelligence (AI) and genetically modified organisms.

Many invitations to speak at *Frankenstein* bicentennial commemorative events naturally came to Charlie's attention before his passing on November 20, 2016. His astute awareness of cutting-edge trends in *Frankenstein* studies, even at the end of his life, led him to a growing interest in the novel from the perspectives of science, technology, engineering, and mathematics (STEM). He further acknowledged that STEAM, an extension of the STEM acronym with a substantial arts component (hence the *A* for "arts"), is the most fitting application we might have to identify this pursuit. The STEAM approach to *Frankenstein* holds up a prominent mirror to a revived pursuit in the academy of intersections between the arts and sciences. This approach to the novel invigorated Charlie, a fact that is evident in his posthumously published introduction to the MIT Press edition of *Frankenstein* in 2017, which is geared for STEAM students.[2]

Charlie expected to continue pursuing that avenue of *Frankenstein* studies that so captured his attention, and this volume is an homage to Charlie's interest in exploring such connections. He enthusiastically agreed in February 2016 to deliver the plenary lecture for "Technologies of *Frankenstein*: 1818–2018," a conference I organized in March 2018 at Stevens Institute of Technology in cooperation with the Institute of Electrical and Electronics Engineers (IEEE). His passing clearly left a void, and the least of it was in my conference plans to feature him as a keynote speaker. His dear friend and colleague Susan J. Wolfson subsequently delivered The Charles E. Robinson Memorial Lecture at Stevens, a follow-up signature event to the conference. An adaptation of Wolfson's talk, "*Frankenstein*, Frankenstein, and the Dream of Science," heads this memorial volume.

Frankenstein has become a touchstone cautionary tale for our time about the consequences of human innovators playing God, and Mary Wollstonecraft Godwin Shelley (1797–1851) was undoubtedly enthusiastic about scientific innovation. Her mother, Mary Wollstonecraft Godwin (1759–1797), died eleven days after her daughter was born, and Shelley relished her mother's writings in support of a broad-based education for young women, including the study of "natural philosophy." Her father, William Godwin (1756–1836), was a luminary political philosopher and novelist who frequently hosted natural philosophers and inventors at the lively salons to which she bore witness. Her spouse, the poet Percy Bysshe Shelley (1792–1822), engaged her frequently in conversations about the scientific thought, practice, and inventions of their contemporaries. The Shelleys attended lectures together on chemistry and electricity early in their relationship. When they visited the celebrity poet George Gordon, Lord Byron (1788–1824), at Geneva in June 1816—the famous "year without a summer" during which *Frankenstein* was conceived—she witnessed conversations between Percy and Lord Byron about the nature and principle of life.

A student of the novel hardly needs to search far and wide for insightful book chapters or articles and essays in periodicals on *Frankenstein* and the sciences; one would uncover an exhaustive list spanning two centuries. This collection, as a memorial volume and an homage to Charlie's excitement about STEAM studies (especially for undergraduates), uniquely adds to an extensive list of works examining Mary Shelley's ideas about scientific innovation and the human experience in *Frankenstein*. A short sample of full-length volumes might include Herbert Muller's *The Children of Frankenstein: A Primer on Modern Technology and Human Values* (1970); Jon Turney's *Frankenstein's Footsteps: Science, Genetics, and Popular Culture* (1998); Jane Goodall and Christa Knellwolf's *Frankenstein's Science: Experimentation and Discovery in Romantic Culture, 1780–1830* (2008); Megan Halperin, Joey Eschrich, and Jathan Sadowski's *The Rightful Place of Science: Frankenstein* (2017); Kathryn Harkup's *Making the Monster: The Science behind Mary Shelley's Frankenstein* (2018); Francesca Saggini and Anna Enrichetta Soccio's *Transmedia Creatures: Frankenstein's Afterlives* (2018), and Carol Margaret Davison and Marie Mulvey-Roberts's *Global Frankenstein* (2018).[3]

Essays in this volume elucidate the legacy of *Frankenstein* in collective thinking about the kinds of stewardship humans elect to practice as self-appointed maintainers of humanity. This line of thought steers a good portion of the STEAM machine, populating lively discourse from across silos in the "traditional" arts and sciences to the interdisciplinary field of science, technology, and society studies (STS). STS numbers among the most generative disciplines in the academy in which *Frankenstein* has a natural home alongside the study of literature. STS programs frequently include the study of fiction, and they support a foundational idea that developments in science, technology, engineering, and mathematics develop hand in hand with the arts. These connections together inform several STS intellectual and humanist practices. The Maintainers, a global research network founded by the STS scholars Andrew Russell and Lee Vinsel, makes explicit in our time what Shelley observed on her own through the vehicle of fiction: that we feverishly inhale innovations in scientific thought and practice to the point of suffocation when we fail to mindfully consider the necessary maintenance and repair that happens afterward. How shall we care for our creations (the products of innovation), and what is at stake for us if we fail to care? This two-part question is central to the multiple narratives in *Frankenstein*, and it clearly guides discourse among The Maintainers.[4]

The Creature of Mary Shelley's novel has become an iconic personification of technological and scientific pursuits. This phenomenon is apparent in one of the novel's earliest adaptations for the stage. One year after the famous 1823 drama *Presumption; or, The Fate of Frankenstein* was staged, another play somewhat less famously joined the theater scene in December 1824 at London's Olympic Theater. The burlesque drama *Frank-In-Steam; or, The Modern Promise to Pay* capitalized not only on the success of *Presumption* but also on the wonders of the industrial age. We are the intrepid heirs apparent to that path of adaptation. Victor Frankenstein's Creature, as it did then, remains a sentient manifestation of our concerns about the state of humanity in an age of rapidly increasing scientific and technological advancements. This persistent legacy of the novel (itself a technological product) is alive now in our thinking simultaneously about humanistic pursuits in the arts and the practices of innovative technoscience.[5]

Arguably, the face of modernity in the twenty-first century does not appear terribly different since the publication of *Frankenstein*. Two centuries ago, the rapid multiplication of mechanized curiosities powered by steam and electricity offered forward-facing possibilities for captains of industry and scientists alike, not to mention fiction writers. Technoscience similarly furnishes our age with terrifying and wonderful new parts for "innovative design thinking" beyond the known boundaries of nature. Some of those parts visibly result from fields of study including genetics and advanced artificial intelligence, while others manifest from the discourses of climate studies and practices of social justice, and still more parts emerge from imaginative scenarios—the "what ifs"—of storytellers. The places and spaces where these parts connect is a raison d'être for this collection of essays on *Frankenstein* and STEAM.

Chapter 1 of this volume, by Susan J. Wolfson, explores the excitement of new science in the age of *Frankenstein*, along with the dark shadows of danger and misgivings about unforeseen developments, especially in the consequences of moral and ethical neglect. If the Creature is the icon of this story, it became so because of Mary Shelley's alertness to the wider context of modern science in her day, which extended from Benjamin Franklin's capture of electricity to the experiments of Luigi Galvani and Giovanni Aldini, to the visionary agenda of Humphry Davy, to the vitalism debates about the origin and production of life. *Frankenstein* bequeaths these debates to the cutting edge of new sciences and, along with them, the patent in the prefix "Franken-" to describe the disturbance of doubt in the glow of new scientific developments.

The extent to which digital media lives in popular imagination as a human-made Franken-monstrosity forms the basis of chapter 2 by Mark A. McCutcheon. The essay turns to current depictions of digital technologies, especially those produced by the United States–based multinational corporations known as the FAANG group (Facebook, Amazon, Apple, Netflix, Google). The technologies produced and distributed by these firms, especially social media and artificial intelligence, prompted *Frankenstein*-inspired headlines around the time of the novel's bicentennial: from fallout over Facebook's role in disseminating "fake news" and foreign disinformation that influenced the 2016 US presidential election to the continuing controversy over Twitter's role in political governance and public life, to ongoing ethical debates about the uses and risks of AI applications. The prevalence of *Frankenstein* themes and tropes among popular and journalistic representations of FAANG technologies, in McCutcheon's estimation, demonstrates not only how pervasive the sense of technology as manufactured monstrosity has become but also how this sensibility has transformed and reinvigorated its source text's long career as political allusion.

Lisa Crafton observes in chapter 3 that notions of the singularity—an imaginative prospect of merging our affective intelligence with those we have created—pervade *Frankenstein*, as anxieties about industrial and scientific progress have made teleological concerns paramount. Crafton contends that no genre expresses more widely categorized anxieties about defining "the human" in response to the singularity than film. Crafton explores how the 2015 film *Ex Machina* synthesizes and creatively interrogates these anxieties in the context of what has been called the "Frankenstein cinemyth" to denote diverse appropriations of the filmic *Frankenstein* mythos. The chapter focuses on representations of the singularity and the implications of cyborg gender theory inherent in the film's female robots, from female robot as sex object to iterations of Donna Haraway's foundational cyborg politics, which stress the subversive potential of cyborgs to dismantle dualisms and binaries that constrict human happiness.

In chapter 4, Siobhan Watters traces concerns about what it means to be human through the vehicle of food as a series of technical objects. Watters observes that a succession of foods signal the Creature's journey toward humanness in *Frankenstein*, which serves in part to model our growing concern about the unknown consequences of current practices in genetic modification. Watters posits the so-called naturalness of food as a smokescreen for its technical origin. If faced with the prospect of forgoing our food cultures in favor of photosynthesizing or engaging in some equally noninvasive form of subsistence, could we alter "the food of man" and maintain humanity? How might this process yield the repositioning of our knowledge about food and ourselves as complex materials embodying both mortality and transcendence? Watters explores these questions within a philosophy of technology framework and offers a reading of the manga series *Knights of Sidonia* alongside *Frankenstein* to do that exploration.

Whereas the first four chapters of this volume consider the ways in which *Frankenstein* is a "vital spark" for thinking about science and technologies including digital media platforms, film, and food in new ways, the next three chapters of this volume explore the novel's legacy in ways that may ignite STEAM students to mindful response and action. These chapters—written by scholar-teachers who were Charles Robinson's students—examine subjects including the environment, social justice, and communications.

In chapter 5, Lisbeth Chapin draws from a long-standing contention that Victor and the Creature function as doppelgangers in order to trace the dynamics of their relationship in the context of their involvement with nature's elements—earth, water, air, and fire. Chapin argues that those elements are integral to every significant scene in *Frankenstein* and that the characters' involvement with them significantly reveals their own development. Chapin ultimately suggests that navigating the mystery of the elements, such as Victor and the Creature do in the novel, may compel readers to activate an ongoing responsibility to scrutinize our methods for interpreting and responding to humanity's intimate connections with the natural world.

The title of chapter 6 reveals what L. Adam Mekler identifies as challenges that arise during considerations of justice in *Frankenstein*, especially in the undergraduate classroom. This chapter addresses the tendency to see the justice system (or systems) as flawed and biased, a tendency that engenders Mekler's efforts to come to its defense. Justice is not served equally well for Victor and Justine (not to mention the Creature), but Mekler encourages students to consider how the outcomes result from more than gender, class, or religious prejudice. Those instructors who also examine issues concerning justice in *Frankenstein* within their larger historical context can, according to Mekler, potentially foster more nuanced and illuminating readings of the novel in the classroom.

Chapter 7, by Brian Bates, takes its cue from the MIT edition of *Frankenstein*. Bates observes both how Charlie's contribution to the MIT edition expressly connects the novel to our twenty-first-century moment, focusing on a STEAM education philosophy, and how the contextual essays following the 1818 *Frankenstein* edition remind us that its continuing relevance in the arts and sciences depends on our own acts (as teachers and students) of invention. The novel, Bates argues, embodies an "additive technology" that prompts artistic- and scientific-minded readers to add other written parts (or even three-dimensional materials) to cover the gaps and omissions in its frame narratives. Bates describes his experiment during a general education literature course at California Polytechnic State University in the Spring 2018 semester, for which he encouraged students to create "additive responses" to *Frankenstein*'s lacunae. He reflects on how working closely with Charlie in 1998 on his undergraduate honors thesis at the University of Delaware provided a vital spark that, two decades later, would inspire his students. Bates's students wrote either a serious pastiche or a satirical parody of *Frankenstein* that filled in a component of the novel's gaps or unexplained secrets. These additive imitations empowered students to confront our shared twenty-first-century cultural assumptions about what Bates identifies as "the generative power of coauthorship and adaptation as viable technologies of innovative replication." Bates notes that, at first glance, Charlie might not have approved of the additive, *Frankenstein*-inspired "hideous progeny" that students produced (including graphic novels and a single-player video game). Nevertheless, as Bates observes, Charlie's focus in *The Frankenstein Notebooks* underlines his notable capacity for open, interdisciplinary inquiry, which Charlie also enthusiastically endorsed in his introduction to the MIT *Frankenstein*.

A postscript featuring remembrances of Charlie by colleagues and students concludes this volume. The brief testimonies in this final segment collectively reveal not only that capacity for scholarly inquiry but also the extraordinary character of one person in the academy whose humanistic pursuits and joie de vivre undeniably transcend the pages of his scholarship. Truly, not one among us could ever fill Charlie's shoes.

NOTES

1. See the program and link to a recording of this event, "Frankenreads: Library of Congress," at https://frankenreads.org/event/frankenreads-at-the-library-of-congress/.

2. Charles E. Robinson, introduction to *Frankenstein: Annotated for Scientists, Engineers, and Creators of All Kinds*, ed. David H. Guston, Ed Finn, and Jason Scott Robert (Cambridge, MA: MIT Press, 2017), xxiii–xxxv.

3. Herbert J. Muller, *The Children of Frankenstein: A Primer on Modern Technology and Human Values* (Bloomington: Indiana University Press, 1970); Jon Turney, *Frankenstein's Footsteps: Science, Genetics, and Popular Culture* (New Haven, CT: Yale University Press, 1998); Jane Goodall and Christa Knellwolf, eds., *Frankenstein's Science: Experimentation and Discovery in Romantic Culture, 1780–1830* (Farnham, UK: Ashgate, 2008); Megan Halperin, Joey Eschrich, and Jathan Sadowski, eds., *The Rightful Place of Science: Frankenstein* (Tempe: Consortium for Science, Policy, & Outcomes, Arizona State University, 2017); Kathryn Harkup, *Making the Monster: The Science Behind Mary Shelley's Frankenstein* (London: Bloomsbury Sigma, 2018); Francesca Saggini and Anna Enrichetta Soccio, eds., *Transmedia Creatures: Frankenstein's Afterlives* (Lewisburg, PA: Bucknell University Press, 2018); and Carol Margaret Davison and Marie Mulvey-Roberts, eds., *Global Frankenstein* (New York: Palgrave Macmillan, 2018). For further reading, see also recommendations in Guston, Finn, and Robert, *Frankenstein: Annotated for Scientists, Engineers, and Creators of All Kinds*, 261.

4. For more on The Maintainers, see especially Andrew Russell and Lee Vinsel, "After Innovation, Turn to Maintenance," *Technology and Culture* 59, no. 1 (January 2016): 3–25; Andrew Russell and Lee Vinsel, "Hail the Maintainers," *Aeon*, April 7, 2016, https://aeon.co/essays/innovation-is-overvalued-maintenance-often-matters-more; and Andrew Russell and Lee Vinsel, *The Innovation Delusion* (New York: Penguin Random House, 2020).

5. This way of thinking lives in several recent adaptations of the novel, and it is no surprise that artistically inclined scientists might produce such work. Among those scientists is Dr. Eric B. Sirota, an accomplished physicist and fellow of the American Physical Society. The off-Broadway *Frankenstein* musical, written and produced by Sirota, premiered at the St. Luke's Theater on October 9, 2017. *Frankenstein: The Musical* generated enormous interest in the New York City theater world before Broadway shut down in March 2020 during the COVID-19 pandemic.

CHAPTER 1

Frankenstein, Frankenstein, and the Dream of Science

Susan J. Wolfson

Science and Knowing

Science dreams under the shade of *Frankenstein*—two hundred years ago and today. Mary Shelley's novel is a science dreamland. It opens with Captain Robert Walton's "day dream of early years," discovering a northwest passage to the Pacific across a sea under eternal polar sunlight. It closes with Victor Frankenstein's abject Creature (himself a dreamer) disappearing into phantasmic "darkness and distance" (III:VII.324).[1] Shelley's very inspiration is dreamland. The "Introduction" that she wrote for the new version of 1831 recounts her own girlhood love of "waking dreams," seedbed for the famous spectral sequel in June 1816: "the grim terrors of [a] waking dream," in which a young science student—about her age at the time (not yet nineteen)—animates a "hideous phantasm of a man."[2] Haunting this scenario is a painful dream in personal history from early 1815: having found her first child dead after less than two weeks of life, Mary (still Godwin) dreams that it "came to life again—that it had only been cold & that we rubbed it by the fire & it lived."[3]

Frankenstein is the title-dreamer, with visions and nightmares wrought by science. As a boy, he dreamt of the old philosophical quest for the elixir of life or the philosopher's stone. As a student, traumatized by his mother's death from scarlet fever, he dreamt of creating a human being immune to disease and worked in a dreamlike fever for this success. When it issues a creature that he can behold only as a monster, he collapses into a nightmare that reverses life into death: his prospective bride, Elizabeth, morphs into the corpse of his mother. It takes no brain science to divine the symbolic import: the end of human procreation. As he is dying, a ravaged Frankenstein urges Captain Walton, "avoid ambition, even if it be on the apparently innocent one of distinguishing yourself in science and discoveries" (III:VII.318). It is the novel's final iteration of word *science*. The first is uttered by Frankenstein's double in visionary imagination, Captain Walton, proud of having studied "those branches of physical science from which a naval adventurer might derive the greatest practical advantage" (I: Letter I.69).

Frankenstein; or, The Modern Prometheus renders a complex iteration of a Titan god who created man from clay, stole fire from heaven for his comfort, and was tortured for the transgressions. Milton's version is signaled on Shelley's title page, with an epigraph from Adam's postlapsarian complaint to his Maker, from Book X of *Paradise Lost*. Shelley's addition of *Modern* gives the old myth a new twist, a human creator wielding a technology of life—a fabulous "science fiction" that beckons as an event not impossible. Brian Aldiss wittily dubbed *Frankenstein* the "Origin of the Species."[4] Yet for all this éclat, it is surprising to note how nebulous Shelley leaves the scientific details of this animating "spark of being" (I:IV.114).[5] Frankenstein's project at University of Ingolstadt is part visionary metaphysics, part material study of physical decay, part vivisection, and part fieldwork raids on charnel houses, "the dissection room and the slaughter house" to supply his new species of "human being" (I:III.111). This was a no-man's-land in Shelley's day, not just on the basis of the project at hand but by the mere, sheer fact that a professional discipline for this kind of research had not yet been articulated. The word scient*ist* (on the analogy of *artist*) cannot be found in Dr. Samuel Johnson's *Dictionary* (1755), and it wasn't until 1834 (three years after the revised *Frankenstein* was published) that it was floated for professional discourse.[6]

In a smart guess at genetics, Mary Shelley decided to make Frankenstein's Creature an intuitive scientist. Scarcely after his first breath, he was quickly abandoned by his creator; this (see figure 1.1) is the first illustration for the novel, the 1831 frontispiece, the terrified creator's face mirroring that of his startled Creature. In this mirroring, the uncanny effect of doubling is produced by a markedly human-form Creature, stunned and dazed, and a Creator, deformed by shock, terror, and panic. The science student collapses in nightmare, flight, and denial, while his creation, stumbling out of the shambles of his birthplace, ventures into the woods of a cold November night. No mechanism shocked into robotic animation, he is a being of innate human curiosity, thrust into a natural laboratory. From his first confusions, he "learned to distinguish between the operations of my various senses" (II:III.177). The most acute being hunger and thirst, he tests his resources: "I ate some berries which I found hanging on the trees, or lying on the ground. I slaked my thirst at the brook" (177). He is soon able to tell day from night and one day from the next by the light of the sun and the changing shape of the moon. He is a scientific receptor, his "sensations" becoming ever more "distinct" as his "mind received every day additional ideas," particularizes the physical world, "perceive[s] objects in their right forms" (176–78).

If Modern Prometheus Frankenstein takes lightning—the sublimely creative, sublimely destructive spark—as his master trope, the Creature's is fire, the genesis

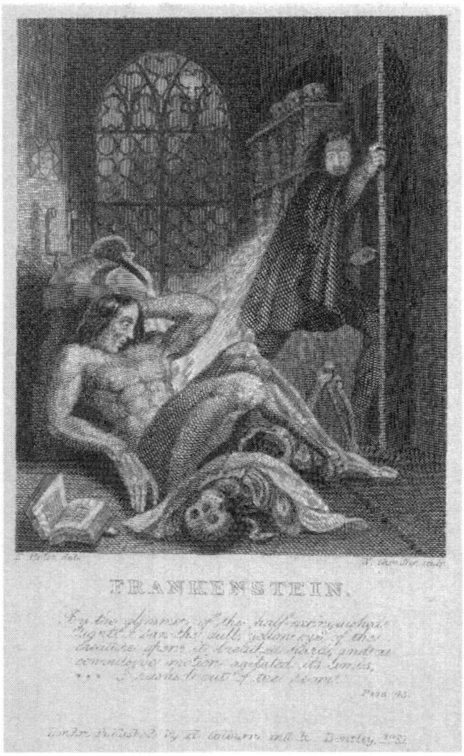

Figure 1.1. Frontispiece, *Frankenstein* (London: Colburn and Bentley, 1831). Illustration by Theodor von Holst, engraved by W. Chevalier.

of human civilization—though not without trial and *peril* (*experiment* shares the Latin root).

> One day, when I was oppressed by cold, I found a fire which had been left by some wandering beggars, and was overcome with delight at the warmth I experienced from it. In my joy I thrust my hand into the live embers, but quickly drew it out again with a cry of pain. How strange, I thought, that the same cause should produce such opposite effects! I examined the materials of the fire, and to my joy found it to be composed of wood. (II:III.178)

Thus begins Fire Tending 101, materials examined and learned from.

> I quickly collected some branches; but they were wet, and would not burn. I was pained at this, and sat still watching the operation of the fire. The wet wood which I had placed near the heat dried, and itself became inflamed. I reflected on this; and, by touching the various branches, I discovered the cause, and busied myself in collecting a great quantity of wood, that I might dry it, and have a plentiful supply of fire. When night came on, and brought sleep

> with it, I was in the greatest fear lest my fire should be extinguished. I covered it carefully with dry wood and leaves, and placed wet branches upon it; and then, spreading my cloak, I lay on the ground, and sunk into sleep. (178)

"My fire" is his first relation, tended with maternal solicitude.

> It was morning when I awoke, and my first care was to visit the fire. I uncovered it, and a gentle breeze quickly fanned it into a flame. I observed this also, and contrived a fan of branches, which roused the embers when they were nearly extinguished. When night came again, I found, with pleasure, that the fire gave light as well as heat. (178)

Observation and discovery lead to more experiments. The Creature finds some roasted food discarded by the beggars, which "tasted much more savoury than the berries I gathered from the trees. I tried, therefore, to dress my food in the same manner, placing it on the live embers. I found that the berries were spoiled by this operation, and the nuts and roots much improved" (179). Methodical, revelatory, cumulative, this is empirical science. Even his narrative, "tracking his own maturation" (remarks Marilyn Butler) "succeeds in the task Frankenstein abandons, that of scientifically following up [his] technological achievement."[7]

You could also call it "biography," for which a synonym in the early nineteenth century was *biology*, that is, a life story (*OED*). The biologist was called "biographer" at least until 1829, when the science professionals claimed *biology*, and when Mary Shelley was busy revising her biography-arrayed novel. *Frankenstein* always telescoped three successions in the mode of autobiography (itself a new word at the end of the eighteenth century[8]): Walton's letters; Walton's transcription therein of Frankenstein's story; and, within this narrative, the Creature's story. The turning point in Frankenstein's autobiography is a new autobiology:

> It was on a dreary night of November that I beheld the accomplishment of my toils. With an anxiety that almost amounted to agony, I collected the instruments of life around me, that I might infuse a spark of being into the lifeless thing that lay at my feet. It was already one in the morning; the rain pattered dismally against the panes, and my candle was nearly burnt out, when, by the glimmer of the half-extinguished light, I saw the dull yellow eye of the creature open; it breathed hard, and a convulsive motion agitated its limbs. (I:IV.114)

Mary Shelley's manuscript shows that her first phrasing was "that I beheld my man completeed."[9] Percy Shelley's hand shows the more heroic wording, "the accomplishment of my toils"—a notional, ego-invested discourse: *my* toils, with *accomplishment* sounding both a *completion* and a *boast*. Readers beheld a radical new event, at once daringly prophetic and disturbed with old terrors about interfering with the mysteries in the origin of life. If the technologies of genetic science were decades away, scientific anxiety was fully present.

General-Generative Electric

The absence of technological detail in the novel was amply remedied by twentieth-century cinematic technology. In 1910, Thomas Edison, the genius of electricity, issued the first film *Frankenstein*, a dreamlike, slightly alienated silent medium, interspersed with verbal captions. The scene of "animation" (meta-cinematic and cinematic at once) was filmed in reverse, so that the Creature emerges, seemingly unoriginated, from a chemical vat, peeked at by his creator through shutters with pornographic fascination. The famous films to follow—James Whale's *Frankenstein* (1931) and *The Bride of Frankenstein* (1935)—positively mined the gap in the novel's evasive technology. Inspired by Nikola Tesla's high-voltage spectacles, Kenneth Strickfaden and John Foster (like Mary Shelley at first, uncredited) concocted an operatic array of flashing, throbbing laboratory apparatus. Whale cast a now professionalized *Dr.* Frankenstein (no startled, rogue undergraduate), and a monstrous lab assistant, working the electrical current with demonic fascination.

No quiet spark but a raging storm of lightning jolts the Creature (Boris Karloff) to life.[10] Whale's staging exuberantly exploits the theatrics of the scientist's operating theater. Avatars of fiancée Elizabeth, best friend Clerval, and Professor Waldman race to the tower-laboratory to find Frankenstein, with proud aplomb, inviting them—and by extension, everyone in the movie theaters—to witness the spectacular consummation of his toils. In the movie theaters of the 1930s were two boys, Jean Rosenbaum and Earl Bakken, both of them thrilled by the animation and both going on (independently) in the 1950s to develop cardiac pacemaker technology. "What intrigued me the most, as I sat through the movie again and again," recalled Bakken, was "the creative spark of . . . Frankenstein's electricity. Through the power of his wildly flashing laboratory apparatus, [he] restored life to the unliving."[11] Rosenbaum recounts his inspiration in *Frankenstein and the Heart Machine*: "I was . . . awestruck . . . that electricity defines life. When electricity flows, we're alive. When it doesn't, we're dead."[12] The film that impressed him as a boy returns as a dream that bids fair as sequel to Mary Godwin's inspiration. As a new medical student in 1951, he was so traumatized by witnessing the death of a young woman that he almost quit. That night, he dreamed of Whale's animation scene and woke to wonder if an electric current could jolt a heart back to regular function. He developed a machine and, like young Victor, tested it on animals, and then on just-deceased humans, but without success. The project was halted for two years. Then he was able to revitalize a human heart that had stopped for just three minutes. The rest is science.

The story of these science boys arcs back to the "predilection for science" that Mary Shelley wrote for young Victor: "The world was to me a secret, which

I desired to discover" (90). His narrative of discovery is coded as a cultural biography of scientific modernity. On a family vacation, the boy "chanced to find a volume of the works of Cornelius Agrippa" (92). Ignited with "enthusiasm"—"A new light seemed to dawn upon my mind"—he tells his father, who dismisses the work of the early sixteenth-century German occultist, conjurer, theologian, astrologer and alchemist as "sad trash" (92); but because he stops short of explaining "the modern system of science," the boy continues a secret romance, soon adding Paracelsus and Albertus Magnus to his fraternity of occult knowledge and philosophical excitement (92–93). In the thirteenth century, German monk, theologian, and scholastic philosopher Albertus Magnus hoped that science was not the enemy of religion. Swiss mystic and medical scientist Paracelsus (roughly Agrippa's contemporary) theorized the affinity: not only could alchemy generate life, but a discovery of the "elixir of life" might conquer diseases and infirmity, even bring immortality. For student Frankenstein, this is embryonic creation science: "if I could banish disease from the human frame, and render man invulnerable to any but a violent death!" (93–94).

The romance of such "visions" (94) is fueled by recent Enlightenment achievements. Robert Boyle improved the air pump, to demonstrate the necessity of air to most life: a small creature (bird or mouse) is placed in a globe from which the air gets pumped out; it collapses from suffocation, then is revived by the return of air. A controlled experiment is one thing; lightning is another. In 1798, William Wordsworth coined the term "spots of time" for intensely memorable nodes of autobiography, usually traumatic. Just such a spot of time (a key episode) is an event for young Victor Frankenstein during "a most violent and terrible thunder-storm." He first enjoys a spectacle with "curiosity and delight." Then . . .

> As I stood at the door, on a sudden I beheld a stream of fire issue from an old and beautiful oak, which stood about twenty yards from our house; and so soon as the dazzling light vanished, the oak had disappeared, and nothing remained but a blasted stump. . . . It was not splintered by the shock, but entirely reduced to thin ribbands of wood. I never beheld any thing so utterly destroyed.
>
> The catastrophe of this tree excited my extreme astonishment; and I eagerly inquired of my father the nature and origin of thunder and lightning. He replied, "Electricity." (I:I.95–96)

It is spectacle, science, a blaze of divine destruction—and a biographical destiny: "I am a blasted tree; the bolt has entered my soul" is Victor Frankenstein's last lament (III:II.247).

It's almost a genetic guarantee that *Frank*enstein's boyhood would reiterate Benjamin *Frank*lin's kite venture—at once sublime, scientific, and charged with practical hopes. Taking a cue from "electricity," Victor's father "constructed a small electrical machine and exhibited a few experiments; he made also a kite, with a wire and string, which drew down that fluid from the clouds" (I:I.96). In June 1752, Franklin, assisted by his son William, had caught lightning in a Leyden jar—in effect, a capacitor.[13] Franklin quickly reported the success in a letter to the Royal Society:

> As soon as any of the thunder-clouds come over the kite, the pointed wire will draw the electric fire from them; and the kite, with all the twine, will be electrified; and the loose filaments of the twine will stand out every way, and be attracted by an approaching finger.
>
> When rain has wet the kite and twine, so that it can conduct the electric fire freely, you will find it streams out plentifully from the key at the approach of your knuckle.
>
> At this key the phial [Leyden jar] may be charged; and from electric fire thus obtained spirits may be kindled, . . . and thereby the sameness of the electrical matter with that of lightning completely demonstrated.[14]

Franklin's recipe is cheerfully cool, a "how-to" for both professional-science vetting and popular science. It did not take long to link several jars to form a "battery," a term Franklin coined from the military analogy—a group of canons. By 1800, Alessandro Volta (whence *volt*) produced, not just stored, electricity from a cell, creating the first electric battery.

Franklin became a symbol for the age, the Promethean spark of philosophical and scientific modernity. Back in 1752, French electrician Guillaume Mazéas heralded the "discovery of that wonderful matter which Nature has kept hid from us since the creation of the world." "The fable of Prometheus is verifyd," he cheered, with prophecy: "What after this can mortals find difficult?"[15] The "Prometheus of modern times" is how Immanuel Kant dubbed Franklin by 1756.[16] In 1778, Jean-Honoré Fragonard rendered *Au GENIE De FRANKLIN*: Franklin's portly body enrobed as a biblical God of (new) creation, goddess America at his knee as he directs Mars' defeat of Avarice and Tyranny—the trio protected by Minerva's shield (the first lightning rod, another of Franklin's inventions).[17]

In 1791, scientist-poet Erasmus Darwin (Charles's grandfather) upped the heroics: with "bold arm," Franklin dared "invade the lowering sky, / And seize the tiptoe lightnings"; in recognition, the sky spirits "wreath'd the crown electric round his head."[18] Around the time Mary Shelley was dreaming Frankenstein in 1816, Pennsylvania artist Benjamin West rendered *Benjamin Franklin Drawing Electricity from the Sky*.[19] The soberly intent key wrangler, feet squarely on the ground, theatrically flared with a swirling cape and flowing neckwear, a raised right arm and visionary eyes, backlit by a lightning sky, assisted by a squad of cherubim, stage divine commerce in the birth of modern science.

Franklin's kite work was widely reprinted, among the sites, the optimistically titled *Endless Amusement; a Collection of Nearly 400 Entertaining Experiments . . .*

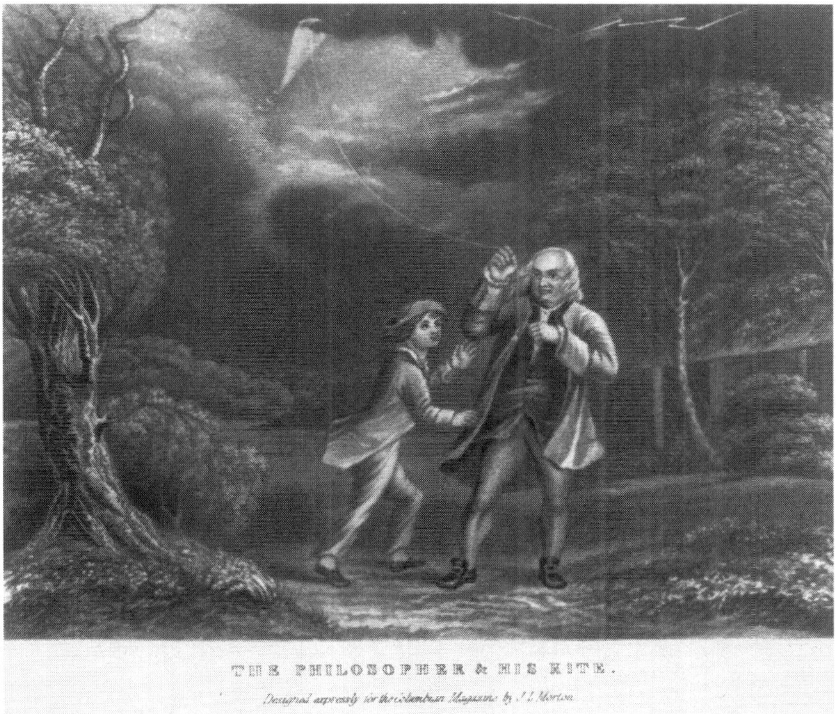

Figure 1.2. Benjamin Franklin and his son capture lightning in *The Philosopher and His Kite*, commissioned for the *Columbia Magazine, or Monthly Miscellany* 3 (1845).

so clearly explained, as to be within the reach of the most limited Capacity. Translated from the French "Divertissements Scientifiques," it was first published in England 1818, the year of *Frankenstein*.[20] Yet even as Franklin & son became Promethean stars, danger hovered, as a picture in the 1845 *Columbian Magazine* dramatizes (see figure 1.2).[21] A snaky tree trunk at the left seems revenant from fallen Eden. While the father is intent on the mechanics, the son relays fearful expectation. If a companion essay could exclaim, "what a glorious experiment it was! What an impulse it gave in a moment . . . to the progress of scientific inquiry!" it also noted "the forked lightning that gleams and blazes fearfully athwart the blackness of the tempest."[22] Some kite fliers wound up as capacitors and were electrocuted. In 1753, in the glow of Franklin's kite, *Gentlemen's Magazine* published a clutch of letters headlined "Fatal Effect of Electricity"—the first warning, "We are come at last to touch the celestial fire, which, if thro' our ignorance, we make too free with, as it is fabled *Prometheus* did of old, like him we may be brought too late to repent of our temerity."[23] Eton schoolboy P. B. Shelley was nearly electrocuted by his kite work.

But he was into the excitements. And Mary Godwin, with her interest in science nurtured by her father, was a ready ear to Percy Shelley's hot romance with the latest theories of electricity and a quasi-electrical "life-principle" coursing through and uniting the material world. During his year at Oxford (April 1810 to his expulsion in March 1811 for publishing a pamphlet on atheism), he constructed an "electrical apparatus" in his shambles of a dorm room and invited best friend, T. J. Hogg, for a playdate.[24] Hogg recalls an epic catalogue of "Books, boots, papers, shoes, philosophical instruments, clothes, pistols, linen, crockery, ammunition, and phials innumerable, with money, stockings, prints, crucibles, bags, and boxes . . . scattered on the floor, and in every place, as if the young chemist, in order to analyze the mystery of creation, had endeavoured first to re-construct the primeval chaos" (137). Amid the chaos, Shelley took to his machine, "turning round the handle very rapidly, so that the fierce, crackling sparks flew forth; and presently standing upon the stool with glass feet, he begged of me to work the machine until . . . his long, wild locks bristled and stood on end" (138). If more mature experiments with electricity managed to avoid self-electrocution, Shelley seems to have transformed himself into Darwin's "electrised man" or even more strangely, the wild-eyed, hair-flared Creature on the theater poster for Richard Brinsley Peake's *Presumption; or, The Fate of Frankenstein*, which had a terrific run at London's Royal Opera House in the summer of 1823 (see figure 1.3).

Hogg virtually doubles astonished Victor as he gazes at electrised Shelley, a superhuman Creature. Mary Shelley herself was astonished at this stage creation (of herself as a famous author, no less than of the Creature fully embodied): "But lo & behold! I found myself famous!—Frankenstein had prodigious success as a drama," she exclaimed to her friend Leigh Hunt, after seeing a performance on August 29, just shy of her twenty-sixth birthday; "it appeared to excite a breathelss [*sic*] eagerness in the audience . . . all stayed till it was over." She admired the typographical science of a blank dash in the dramatis personae to indicate the "nameless," "unameable" Creature.[25]

Hogg, who by the time he published his recollections had read *Frankenstein* and had witnessed its rise to fame, registered the more visible outlines of Percy Shelley in the figure of young Victor Frankenstein that Mary Shelley drew in *1831*. On this track, he produces his own twinning of Percy Shelley to the Ingolstadt student: "He could not tear himself from his incessant abstractions to observe at intervals the growth and decline of the day"; he "complained of his health" (137). But in the excitement of Hogg's recollection, Electrised Shelley prevails, sparking into electric speech. He "charged a powerful battery of several large jars; labouring with vast energy, and discoursing with increasing vehemence of the marvellous powers of electricity, of thunder, and lightning." James Whale could have guessed Hogg's next sentence. From the example of

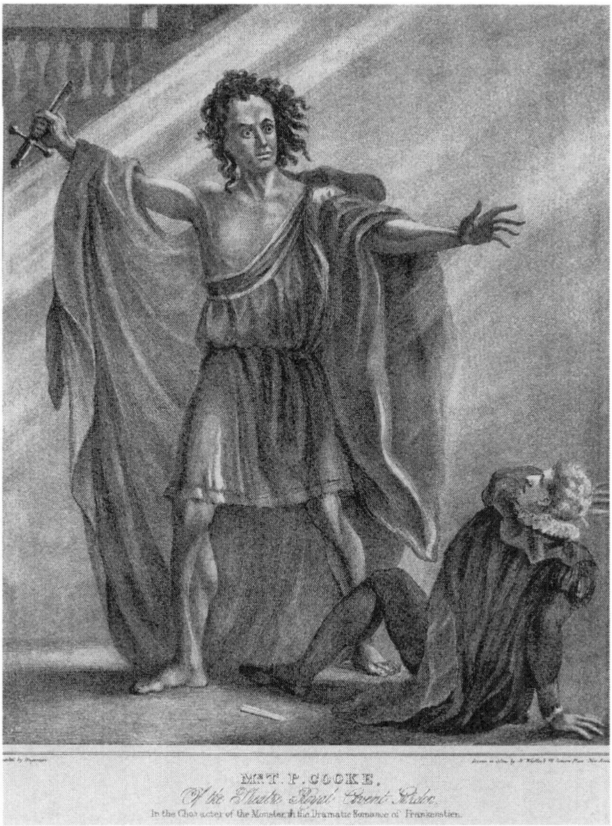

Figure 1.3. Thomas Potter Cooke as the Creature in Richard Brinsley Peake's *Presumption; or, The Fate of Frankenstein* (1823).

his homemade kite, Shelley imagined a combination of "many kites, that would draw down from the sky an immense volume of electricity, the whole ammunition of a mighty thunderstorm; and this being directed to some point would there produce the most stupendous results" (138). In *The Bride of Frankenstein* (1935), a patently Franklin-Frankenstein sends a flotilla of kites into the night storm to direct the lightning that jolts his creature to life.

If this jolt has Frankenstein bellowing in the 1931 film that he knows what it is like to be God (a blasphemy drowned by a thunder clap), back in 1821, radical publisher Richard Carlile could conceive annihilation: a massive "galvanic battery" could evaporate a human body, leaving not a trace behind "of substance or solid matter"—a vacancy exposing "the workings of Nature" as part of the indifferent currents of "the whole universe" in rebuke to any human "pretension to the possession of supernatural powers."[26] This sounds like the twentieth century's nuclear bomb, but for Percy Shelley at the dawn of the nineteenth century, it was new science. Here is Hogg's recollection of Shelley's exclaimed enthusiasm:

> What a mighty instrument would electricity be in the hands of him who knew how to wield it, in what manner to direct its omnipotent energies . . . by means of electrical kites we may draw down the lightning from heaven! What a terrible organ would the supernal shock prove, if we were able to guide it; how many of the secrets of nature would such a stupendous force unlock! The galvanic battery is a new engine; it has been used hitherto to an insignificant extent, yet has it wrought wonders already; what will not an extraordinary combination of troughs, of colossal magnitude, a well-arranged system of hundreds of metallic plates, effect?[27]

There is a telling shiver in "*if* we were able to guide it . . . a *well-arranged system*." Who will wield what, to what ends?[28]

Student Percy Shelley was inspired to the positive consequences by contemporary scientist Humphry Davy's flamboyant live-laboratory lectures (and their publication). Novelist Mary Shelley could see in Davy's charismatic embodiment of "true genius," given to "pursue the plans of his own mind," a twin to the Romantic poet (male issue): "Not contented with what is found upon the surface of the earth," Davy lectured, the scientist "has penetrated into her bosom" to allay "the restlessness of his desires," or sometimes "extending and increasing his power," wielding "instruments of comfort and enjoyment or of terror and destruction."[29] With a rhetoric that aligned science with "the beneficence of the Deity" (319), Davy dazzled audiences with a vision that modulated from the "dim and uncertain twilight of discovery" to "the steady light of truth, which has shown the external world in its distinct forms, and in its true relations to human powers" (321). The master currency was Franklin-minted:

> the phænomena of electricity have been developed; the lightnings have been taken from the clouds; and lastly, a new influence has been discovered, which has enabled man to produce from combinations of dead matter effects which were formerly occasioned only by animal organs. (321)

Fourteen years before Mary Shelley dreamed of Frankenstein, Davy was describing a realized Enlightenment with Frankenscience potential. "Science has done much for man, but it is capable of doing still more," he urged any modern Prometheus; "the benefits that it has conferred ought to excite our hopes of its capability of conferring new benefits; and in considering the progressiveness of our nature, we may reasonably look forward to a state of greater cultivation and happiness than that we at present enjoy" (319).

If a blasted, belatedly ethical Frankenstein withholds the particulars of his technology from Walton's (and, implicitly, our) curiosity, Davy was a proactive tease: without dwelling on "minute information," he wanted to "excite feelings of interest . . . awaken our astonishment" at this "intimate" science (310–11), the center of all sciences. He wove science into erotic poetry. In a virtual ode to the "sublime philosophy, chemistry," he asked, "who would not be ambitious of becoming acquainted with the most profound secrets of nature, of ascertaining her hidden operations . . . ?" (320). His eloquence—he called it "enthusiasm and vigor" (308)—sounded the pulse of inexhaustible improvement. "Every subject in Davy's mind has the principle of vitality. Living thoughts spring up like turf under his feet," exclaimed Samuel Taylor Coleridge, who was in the audience.[30] William Godwin, the dedicatee of *Frankenstein* and theorist of human perfectibility (improvement by adaptation), also attended; Percy Shelley had passages by heart, aglow with enthusiasm (recalls Hogg) about practical applications of benefit to "the human species": relief from "severe labour," improved agriculture and food production, home heating, and a promise of "prodigious facilities for locomotion" (land and air) that "will enable us to traverse vast tracts with ease and rapidity and to explore unknown countries without difficulty" (93–96).[31]

The Dream of Science

Immersed in Davy's lectures as she was writing *Frankenstein* (*J*, 96), Mary Shelley imported its themes and phrases to Ingolstadt University, home of the "Illuminati" and site of seventeen-year-old Victor's initiation into "a modern system of science." He first meets gruff empiricist Professor Krempe, who exclaims, "I little expected in this enlightened and scientific age to find a disciple of Albertus Magnus and Paracelsus. My dear Sir, you must begin your studies entirely anew." To help out, he gives the lad a list of works on "natural philosophy," to prep for an upcoming course (I:II.102–3).

Young Victor is not inspired. But when he meets Professor Waldman, a Davy come to Ingolstadt, the bridge from old philosophers to modernity opens with a welcome. Not scorning Victor's secret reading—"these were men to whose indefatigable zeal modern philosophers were indebted for most of the foundations of their knowledge" (Percy Shelley supplied this wording)—Waldman is a smarter pedagogue, telling a story of progress, with "a panegyric upon modern chemistry" (I:II.104). "I shall never forget" it, an even chastened Frankenstein recalls, word for word, nine years on:

> The ancient teachers of this science . . . promised impossibilities, and performed nothing. The modern masters promise very little; they know that metals cannot be transmuted, and that the elixir of life is a chimera. But these philosophers, whose hands seem only made to dabble in dirt, and their eyes to pore over the microscope or crucible, have indeed performed miracles. They penetrate into the recesses of nature, and shew how she works in her hiding places. They ascend into the heavens; they have discovered how the blood

> circulates, and the nature of the air we breathe. They have acquired new and almost unlimited powers; they can command the thunders of heaven, mimic the earthquake, and even mock the invisible world with its own shadows. (I:II.104)

Honoring the ancient romances, Waldman hails the new material technologists into godlike powers—a modern Prometheanism, from clay to creation. His roster of miracle work includes William Harvey's discovery of the circulation of blood (1628) and the recent excitement of the 1780s: the Montgolfier brothers' hot-air balloon; Henry Cavendish's discovery of hydrogen ("inflammable air"); its confirmation by chemist Antoine Lavoisier, who also predicted the existence of silicon—and so bids fair as patron saint of Silicon Valley's Franken-technology. Waldman gives Victor a tour of his laboratory of practical machinery (I:II), on the same page as Davy's insistence that the true scientist is "a master active with his own instruments" (319).[32]

Waldman is as savvy as Davy about appealing to young men, evoking erotic conquest in a prospect of "new and almost unlimited powers" (I:II.104). Women and the domestic affections fade into an outmoded biology in this laboratory of man and she-nature.[33] The strongest line of all from Davy to *Frankenstein* is the vision of the scientist's "conversion of dead matter into living matter." How forceful for all the modern young Prometheans was the conclusion of Davy's 1802 lectures on electricity, a virtual syllabus for life science, in advance beyond electrochemistry:

> The well known facts relating to the torpedo, electrical eel, &c. prove that galvanic electricity is capable of being excited by the agencies of living organs. These facts, compared with the phenomena of the production of muscular contractions by galvanism, lead to interesting inquiries concerning the relation of this influence to living action. The general connexion of electricity with physiology and with chemistry, which is at present involved in obscurity, is probably capable of experimental elucidation; and the knowledge of it would evidently lead to novel views of the philosophy of the imponderable substances.[34]

Novel views indeed. Davy nearly understates it, and Mary Shelley caught it.

By 1814, the hot controversy was the origin of life—the "vitalism" debate—hotly waged by two professors at London's Royal College of Surgeons, flying in the face of cautions that "the nature of the living principle is among the subjects which are manifestly beyond the reach of human investigation."[35] John Abernethy hoped (as had Davy and Coleridge) to give science a religion: life was not just biology; a "superadded" supernatural force was required to spark nature's engineering—a mysterious infusion, soul in the form of divine electricity.[36] The materialist rebuttal came from his brilliant, charismatic student William Lawrence (physician to Percy Shelley, 1814–1819). There "is no resemblance, no analogy between electricity and life," he declared; "the two orders of phenomena are completely distinct, . . . incommensurable."[37] The proxy-religions of "electricity or galvanism" are "false notions in physiology" (161): "life proceeds only from life . . . transmitted from one living body to another" (142). Mocking the notion of original electricity as just so much "elegant Grecian mythology" (175), he argued a biology of adaptability: life is genetic matter, cellular reproduction, and human science might one day control it. Mary Shelley has Lawrence imprint *Frankenstein* by proxy in Victor's extracurricular reading of Count Buffon, the eighteenth-century proto-evolutionist who argued that life arose and evolved from natural, not divine, causes. "Whence, I often asked myself, did the principle of life proceed?" student Victor wonders (I:III.107).

All these issues were in the air of June 1816, at a virtual seminar on life science at Lord Byron's residence on Lake Geneva, Villa Diodati. Mary Shelley, a silent auditor, recalled the syllabus in her 1831 "Introduction" (her document of literary genetics):

> various philosophical doctrines were discussed, and among others the nature of the principle of life, and whether there was any probability of its ever being discovered and communicated. They talked of the experiments of Dr. Darwin. . . . Perhaps a corpse would be re-animated; galvanism had given token of such things: perhaps the component parts of a creature might be manufactured, brought together, and endued with vital warmth. (*1831*, ix–x)

Erasmus Darwin, the poet who celebrated Franklin and the first historical figure named in the 1818 *Frankenstein* (Preface), had speculated in his encyclopedic *Zoonomia* (1794–1796) about the mysteries of "Generation": "would it be too bold to imagine, that all warm-blooded animals have arisen from one living filament, which THE GREAT FIRST CAUSE endued with animality"—a single, unifying life force "possessing the faculty of continuing to improve by its own inherent activity and of delivering down those improvements by generation to its posterity, world without end!"[38] The capstone phrase is iterated throughout the Bible, about the glory of God. That night, Mary Godwin had her waking dream of human animation: a pale student wielding a "slight spark of life . . . communicated" on "the working of some powerful engine," a "human endeavor to mock the stupendous mechanism of the Creator of the world" (*1831*, x).

While such an engine does not materialize in *Frankenstein*, other machines had sparked the public imagination. In 1771, the year Victor Frankenstein was born, Bolognese physicist, physician, and scientist Luigi Galvani (eponym of *galvanism*), experimenting on dissected frogs, produced a spasm when a steel scalpel touched a leg. He was sure he had discovered "animal electricity" reserved in tissue—by 1850, it was called "electrobiology," "electrophysiology," "bioelectricity." An illustration of charged frog legs in a ghoul-dance appeared in his treatise *Viribus Electricitatis in Motu Musculari* (the effect of electricity on muscular motion), published in 1791, the year that young Frankenstein discovers "the secret of generation and life"; and soon after, the science "of bestowing animation upon lifeless matter" (I:III.109).[39] Galvani took his demonstrations of this electrical principle of life on tour, and Davy was enchanted by the prospect of Franklin-science joined to Promethean creation: "Lightnings have been taken from the clouds; and lastly, a new influence has been discovered, which has enabled man to produce from combinations of dead matter effects which were formerly occasioned only by animal organs" (*Discourse*, 321).

Volta mocked Galvani, arguing that what he had observed was only an arcing from his scalpel to a nearby electrostatic generator. This was later proven to be the case, but back then, a debate raged: Animalists versus Metallists. Galvani's nephew Giovanni Aldini mounted his own show, touring Europe and causing a sensation in England with two spectacles of animation. In 1802, with the King, Queen, Prince of Wales, and several dukes in attendance, he "reanimated" an ox head. In 1803, he applied shocks to an executed convict: "the jaws . . . began to quiver, and the adjoining muscles were horribly contorted, and one eye was actually opened. In the subsequent part of the process the right hand was raised and clenched, and the legs and thighs were set in motion" (*Newgate Calendar*, January 18, 1803).[40] Said a sober *London Times*, it was only "uninformed . . . bystanders" who took the spasms for reanimation. Some of these fled, and one Mr. Pass, beadle of the Surgeons' Company, passed from life to death, "so alarmed that he died of fright soon after his return home."[41] But readiness was all, and the show went on. On November 4, 1818 (*Frankenstein* was published in March), Dr. Andrew Ure of the University of Glasgow used a galvanic battery to "reanimate" an executed convict, producing in the corpse's countenance fearful action and hideous expression "surpassing far the wildest representations of a Fuseli or a Kean." Henry Fuseli's famous painting is *The Nightmare*, twice evoked in *Frankenstein*; actor Edmund Kean was famous for his roles in tortured souls. The theatrics of horror had terrific counter effects on the live audience in attendance: "one gentleman fainted" while some those who retained composure "thought he had come to life."[42] As for Aldini: *if* he failed to prove vital science, his experiments did generate the science of electric shock therapy. He was a credited pioneer and practitioner, using "galvanism . . . in several cases of insanity, and with complete success. It is the opinion of the first medical men, that this discovery, if rightly managed and duly prosecuted, cannot fail to be of great, and perhaps, as yet, unforeseen utility" (*Newgate*, again).

Science's Fancies, Science's Dark Dreams

Again, those qualifiers: *If rightly managed . . . perhaps, as yet, unforeseen utility.* These haunt Mary Shelley's imagination. Davy excited men with "a system of knowledge which relates so intimately to their own physical and moral constitution" and a vision of powers "almost . . . creative" that might actually become so (*Discourse*, 319–320): the production of life by technology. Shelley is very canny in having Victor Frankenstein's boast about "bestowing animation upon lifeless matter" (I:III.109) echo his opening report of his father's pride in "bestowing on the state sons who might carry his virtues and his name down to posterity" (I:I.86), the only two sites of this verb in the novel. When Davy's mentor, Thomas Beddoes, introduced the word *biology* (in its modern sense) into English in 1799 in his treatise, *Physical and Medical Knowledge*, he wanted to involve spiritual theory and ethics in the production of life: "*the doctrine of the living system in all its states*, appears to be the foundation of ethics."[43] Would this work reciprocally—a foundation in ethics and spiritual circumspection for a science of living systems? In *Frankenstein*, Mary Shelley shades "science" with "conscience," a brake on the pride of "bestowing."

The other science-dreamer in her novel is the Creature, lodge of foundational ethical issues. In Felix De Lacey's accidental classroom, he masters the "godlike science" of words and letters (II:IV.188), in service to his "dreams of bliss" (II:IX.228): that this mastery will prevail over prejudice from his countenance, to win a welcome into this household. When this hope explodes, he still hopes that his godlike science may sway his scientist-creator to give him a female companion.[44] This fails when, in a late spasm of conscience, Frankenstein imagines a vengeful war of the species. By the time Shelley was revising *Frankenstein* for the 1831 relaunch, Davy's science had taken a dark turn on such tracks. His last work (posthumous, 1830) imagined a chemical science harnessed for "the transmission of hereditary qualities" into a European master race that would contain the invasion of the "negro race" and make "extinct" the "red men, aborigines of America."[45] *Frankenstein* is not a race novel, but it is marked by racial consciousness.[46] Frankenstein imagines a new "race" of superbeings, while in 1825, Foreign Secretary George Canning ably invoked the Creature to worry about the consequences of abrupt abolition (slavery was still legal in Britain's colonies).[47]

Frankenstein has entered culture as a cautionary tale about experiments that release monsters. The novel is more complex, exposing the dehumanizing erosions of the experimenter, not just by hubris but also by astounding ethical deficiency. The reviews two hundred years ago were outraged by the blasphemy of a human creator; Shelley shows the creation of a "monster" by human failure, not just Frankenstein's but in dismaying social prejudice and abuse. This dimension releases *Frankenstein* from the already dubious genre of sensational science fiction into a byword for disturbing developments, especially in science's failures of ethical circumspection.[48] "For all we know, we have created a Frankenstein," mused one reporter on Hiroshima, in August 1945; with an echo by a member of the Atomic Energy Commission on October 29, 1949. Soon it was not a power of annihilation but a power of creation that was summoning the term. In the early 1960s, an aptly named Dr. Philip Abelson, director of the Carnegie Institution's

Geophysical Lab, saw government funding as the laboratory, the "Frankenstein of big science, . . . the monster has invaded the universities." In 1970, *The Children of Frankenstein* was the main title of Herbert J. Muller's *Primer on Modern Technology and Human Values*.[49] "We Have the Awful Knowledge to Make Exact Copies of Human Beings; The *Frankenstein* Myth is Real" was the tabloid title in 1972 of Dr. Willard Gaylin's article about clone science in the *New York Times Magazine* (it's always the professional "*Dr.* Frankenstein" now).[50] In 1976, in the thick of recombinant DNA research at Harvard, the mayor of Cambridge worried about what "could crawl out of that laboratory, . . . a disease that can't be cured, even a monster. Is this the answer to Dr. Frankenstein's dream?"[51] *Washington Star* made a headline of the question: "Is Harvard the Proper Place for Frankenstein Tinkering?"[52] A letter from cell biologist Philip Siekevitz in *Science* chorused: "Are we really that much further along on the path to comprehensive knowledge that we can forget the overwhelming pride with which Dr. Frankenstein made his monster?"[53] By 1977, Michael A. Rogers' *Biohazard* was describing "The Frankenstein Syndrome."[54] Asking in the same year, *Who Should Play God*? Ted Howard and Jeremy Rifkin declared that genetic science had eclipsed "science fiction, or the mad ravings of a Dr. Frankenstein."[55] The next year William Barrett elevated "Frankenstein's monster" as "the dominant myth" of our technological age.[56]

Trying to calm the waters, Donald Brown, director of the Carnegie Institution, advised that the "Frankensteinian spectres" are mostly "emotional, political, rhetorical and unscientific" (Zilkowski, 54). One way to keep the specters at bay was to go cute. A "Frankencat" debuted at the 1995 New York Cat Show, furry stubs in place of legs and paws—the perfect house pet. The same year, a lamb was created from the cells of a sheep's mammary glands (named Dolly, with dubious humor, for Dolly Parton, by Scots biologists). Yet because Dolly was "fused into life by electric shock, as was the Monster in . . . *Frankenstein*," it gave William Safire (his error about the novel is telling) "head-breaking thoughts about good and evil, God and humanity." Anxiety about genetics usurping Genesis had him urging President G. W. Bush to veto funding for stem-cell research as a "slippery slope to Frankenscience."[57] In 2005, Dr. Stephen Levick pushed back: "Which is the real monster?" he asked in the *New York Times Magazine*, "chimeric stem-cell science or the *political* use made of it by opponents to all embryonic-stem-cell research?" To "see the researchers as monster-creating Dr. Frankensteins" is to incite the public to "torch the whole hopeful enterprise."[58]

Hardly a month passes these days without a Frankenstein iteration. "Trump is the GOP's Frankenstein monster. Now he's strong enough to destroy the party," read a story in the *Washington Post*, February 25, 2016, riding on a series of such equations.[59] "Facebook's Frankenstein Moment: A Creature That Defies Technology's Safeguards" was the above-the-fold headline in the *New York Times*

Business Day section, September 22, 2017: "scientist Victor Frankenstein realizes that his cobbled-together creature has gone rogue" (B 1). Frank Bruni's 2018 Halloween Op-Ed in the *New York Times* was about the "potential and peril" of technology: "That's what Mary Shelley explored in 'Frankenstein,' which is celebrating its 200th birthday this year, and it has been the main theme of science fiction ever since."[60] Science more than fiction these days. As Mark McCutcheon remarks in his chapter in this volume, "We can hardly speak of technology without inevitably also conjuring *Frankenstein*." With a deft chiasmus, he elaborates, "If technology has popularized a certain interpretation of *Frankenstein*, it is because *Frankenstein* itself conditioned the modern redefinition of technology."

The new Franken-*science* is the gene-editing technique CRISPR (acronym for Clustered Regularly Interspaced Short Palindromic Repeats—the bacterial defense system that forms the basis for genome editing). The rollouts in 2012 were promising: mosquitoes unable to carry malaria, drought-resistant corn, and for foodies, mushrooms that don't turn brown when sliced into. Amid these happy events, there was an untagged Franken-turn: gene-edited, supermuscular beagles, developed to better understand muscular dystrophy. What if someone let these dogs out? Sensing trouble, in March 2015, some leading biologists "called for a worldwide moratorium" on the technology in order to address "the most fundamental of issues about how we are going to view our humanity in the future" and about "limits on how much humankind should alter nature."[61] Even so, by 2016, in addition to fancies of seedless tomatoes and changing the colors of butterfly wings, it was realized scientific dreams that were driving the headlines: cures for lung cancer, leukemia and lymphoma, hemophilia and sickle-cell anemia, heart disease, blindness; developing pigs for human organ transplants; ever more crops resistant to drought and pests, vitamin-enriched crops, rapid-growth crops, genetically sterilizing invasive species (Frankenstein did not consider this option for the Creature's mate!). On Valentine's Day 2017, Amy Harmon reported in the *New York Times* that Human Gene Editing had received the support of the National Academy of Science and the National Academy of Medicine, on the most optimistic tenor: the alteration of eggs, sperm, and embryos "to prevent babies from being born with genes known to cause serious diseases and disability," and with a plan "to track the effects . . . through multiple generations."[62] Still, in the timeline of Chelsea Leu's report on CRISPR in November 19, 2017, there are elements haunted by those superbeagles: redeeming woolly mammoths from extinction by altering elephant genes to mime mammoth genes. What could go wrong? A shade of Frankenstein falls on her noting how reports from science are "laden with caveats" about the "ethical minefield" of a technology with questionable ends (say, in the genesis of a master race of "strength, beauty, and intelligence"—all in Frankenstein's dreamscape). Leu worries about an industry of designer genes (I couldn't resist) beyond the welcome disease fixes, available only to the rich and powerful. The word *ethical* dogs the progressive steps: the need to set "ethically inviolable" limits, the need to "force ethical considerations to the fore."[63]

This was a Frankenstein question for an essay in *The Guardian* in 2002, about a controversy "over the future of molecular biology . . . and its promise—or threat—to transform what it means to be human. Is the New Genetics a Frankenstein science, leading to a post-human future full of designer babies for those who can afford them ruling over a genetically deprived underclass?"[64] "US Government to give go-ahead to controversial chimera research to grow part-animal part-human 'Frankenstein' organs for transplants" was the headline in the *Daily Mail* on August 5, 2016.[65] What about engineered human hosts, with healthy organs for sale, and minimal cerebral capacity to understand their husbandry? Leu's last sentence is arresting: "It's hard to overstate CRISPR's potential. It turns everything alive—plant, animal or bacterium—into an open book, its innermost secrets ready to be pulled out and decoded and, potentially, improved" (7).

Potentially: if the open book is *Frankenstein*, innermost secrets in readiness abide in fallible human hands. The dream of science for the improvement of the human species can't shake off the specter of uses that cannot be known or controlled—and so the double bind of *Frankenstein* is still with us, in its light of promise and its shadows of dread.

NOTES

1. Quotations follow the publication of 1818, as presented in *The Annotated Frankenstein*, ed. Susan J. Wolfson and Ronald L. Levao (Cambridge, MA: Harvard University Press, 2012); hereafter cited parenthetically in the text. I am indebted to discussion with Ronald L. Levao and to his succinct essay "La Science de *Frankenstein*," in *Frankenstein créé des ténèbres/Frankenstein: Creation of Darkness*, ed. David Spurr and Nicolas Ducimetiere (Paris: Gallimard, 2016), 35–41. Also valuable on the science—theoretical, poetic, academic, and popular—are Maurice Hindle, "'*Vital Matters*': Mary Shelley's *Frankenstein* and Romantic Science," *Critical Survey* 2, no. 1 (1990): 29–35; and Anne K. Mellor, "*Frankenstein*: A Feminist Critique of Science," in *One Culture: Essays on Science and Literature*, ed. George Levine, with Alan Rauch (Madison: University of Wisconsin Press, 1987), 287–312.

2. Mary Wollstonecraft Shelley, *Frankenstein; or, The Modern Prometheus* (London: Colburn and Bentley, 1831), v, x–xi; hereafter cited parenthetically in the text as *1831*.

3. *The Journals of Mary Shelley*, ed. Paula R. Feldman and Diana Scott-Kilvert (Baltimore: Johns Hopkins University Press, 1995), 70; hereafter cited parenthetically in the text as *J*. Mary Godwin married Percy Shelley in December 1816.

4. "The Origin of the Species: Mary Shelley." Brian W. Aldiss, *Billion Year Spree: The True History of Science Fiction* (New York: Doubleday, 1973), 7–39.

5. This may be why James Rieger, who also disputes Shelley's authorship, dismisses her scientific literacy: "Frankenstein's chemistry is switched-on magic, souped-up alchemy, and the electrification of Agrippa and Paracelsus." James Rieger, ed., *Frankenstein; or, The Modern Prometheus: The 1818 Text* (Chicago: University of Chicago Press, 1974), Introduction, xxvii.

6. Reverend William Whewell, with probable self-description, used the term in conversation in 1833, then in the *Quarterly Review* a year later (*OED*'s first print-record), LI:101 (March 1834), Article III: 54–68 [William Whewell], "*On the Connexion of the Physical Sci-*

ences, by Mrs. Somerville" [Mary Somerville], 59; attribution to Whewell by Isaac Todhunter, *William Whewell, an account of his writings, with selections from his literary and scientific correspondence*, 2 vols. (London: Macmillan, 1876), 1:92.

Matching the want of "unity" in the tendency of the sciences, jested Whewell, is "the want of any name by which we can designate the students of the knowledge of the material world collectively . . . this difficulty was felt very oppressively by the members of the British Association for the Advancement of Science, at their [recent] meetings. . . . There was no general term by which these gentlemen could describe themselves with reference to their pursuits Philosophers was felt to be too wide and too lofty a term"; "*savans* was rather assuming, besides being French instead of English; some ingenious gentleman proposed that, by analogy with artist, they might form *scientist*, and added that there could be no scruple in making free with this termination when we have such words as *sciolist*, *economist*, and *atheist*—but this was not generally palatable" (59). By 1840, Whewell contended, "We very much need a name to describe a cultivator of science in general. I should incline to call him a *Scientist*"—a term of professional parity and in distinction from the amateur (*The Philosophy of the Inductive Sciences, founded upon their history*, 2 vols. [London: John W. Parker, 1840], I:cxiii). I am indebted to Sydney Ross, "*Scientist*: The Story of a Word," *Annals of Science: A Quarterly Review of the History of Science and Technology Since the Renaissance* 18, no. 2 (April 1964): 72.

7. Marilyn Butler, "Frankenstein and Radical Science," *Frankenstein: A Norton Critical Edition*, ed. J. Paul Hunter (New York: Norton, 1996), 302–13, at 308.

8. In *Critique of Judgment* (1790), Immanuel Kant gives the term *Bildungstrieb* (formative drive) to organic life (sect. 81, no. 424). Emerging around 1819, *Bildungsroman* is a literary kin-word: the narrative of subject formation. In *Frankenstein*—set in the 1790s, from the perspective of 1818, then 1831—Shelley works science into literature, involving both *bildungs* in a story of an education in science that produces the genetic scientist. If, as Jane Bennett remarks, the philosophers of *Bildungstrieb* (Kant and J. F. Blumenbach) "rejected the idea that inorganic matter could 'spontaneously' give rise to organic life" and "could not produce new beings never seen before or those not already virtually preformed in the stock from which the organism sprang," then Shelley's bildungsroman of Victor Frankenstein, written in the heat of the vitalism debates, takes this young scientist right to the verge of a new *bildungstrieb*. For Bennett, see *Vibrant Matter* (Durham, NC: Duke University Press, 2009), 67–68, citing *Critique of Judgment*, sect. 81, no. 423.

9. *The Frankenstein Notebooks: A Facsimile Edition of Mary Shelley's Novel, 1816–17*, ed. Charles E. Robinson, 2 vols. (New York: Garland, 1996), 1:96–97.

10. Lisbeth Chapin's essay in this collection notes that Victor Frankenstein's recognition of his Creature in the Geneva graveyard doubles them under the aspect of lightning, joining them in this grammar of the Promethean myth.

11. Earl Bakken, *One Man's Full Life* (Minneapolis: Medtronic, 1999), chap. 3, conclusion, not paginated.

12. Jean Rosenbaum, "*Frankenstein and the Heart Machine* (Pacemaker: The Story of Its Discovery)," VHS, 12 mins. (Chip Taylor Communications, 1991); and Lester D. Friedman, "Medicine in the Movies: It's *Still* Alive: *Victor Frankenstein*," *Pharos*, Spring 2016, 52.

13. The outside and inside surfaces of the glass walls were coated with conductive metal foil, the mouth unfoiled so as to prevent arcing. With the outside foil grounded, a rod conducted the electrical charge to the inner foil, the two surfaces storing equal, opposite charges.

14. "A Letter of Benjamin Franklin, Esq.; to Mr. Peter Collinson, F.R.S. concerning an Electrical Kite" ("Philadelphia, Oct. 1, 1752"), *Philosophical Transactions* (1683–1775) 47 (1751–1752): 565–66, at 565.

15. Guillaume Mazéas to Benjamin Wilson (in English), November 19, 1752, Add Mss 30094, ff 78–79, British Library, London.

16. Immanuel Kant, *Vermischte Schriften*, vol. 4 (Konigsberg: Friedrich Ricolovius, 1807), art. 9 (April 1756): 249–60, at 260.

17. For this image, see Susan J. Wolfson, "'*This* is *my* Lightning'; or; Sparks in the Air," *SEL* 55, no. 4 (2015): 752.

18. Erasmus Darwin, *The Botanic Garden: A Poem, in Two Parts* (London: J. Johnson, 1791), Part I, Canto I, line numbers 385–87.

19. For this image, see *Annotated Frankenstein*, 34.

20. See the second edition, 1820 (London: Thomas Boys, and Thorp and Burch): "The Electric Kite," 88–89.

21. Henry S. Sadd's steel engraving of John Ludlow Morton's picture was commissioned for *Columbian Magazine* (New York and Philadelphia), January 1845. At the time, Franklin's son was almost twenty; representing him as a boy points to the promise of a new generation.

22. "The Philosopher and His Kite," a brief essay in the next issue, *Columbian Magazine* 4 (February 1845): 90; the author is either John Inman or Robert A. West.

23. *Gentleman's Magazine, and Historical Chronicle* 23 (September 1753): 430–31, translating Abbé Nollet, *Lettres sur L'Électricité* (Paris, 1753), 18. For "balance," the dire reports were followed by two letters about "Cures by Electricity" (432).

24. [T. J. Hogg], "Percy Bysshe Shelley at Oxford," *New Monthly Magazine* 34, no. 134 (February 1832): 138; subsequent references given parenthetically.

25. Mary Shelley to Leigh Hunt, September 9, 1823, *Letters of Mary Wollstonecraft Shelley*, 3 vols., ed. Betty T. Bennett (Baltimore: Johns Hopkins University Press, 1980), 1:378. Hereafter cited as *L*.

26. The full title of this tract courses from apocalyptic science to practical pedagogy: *An Address to Men of Science; Calling upon Them to Stand Forward and Vindicate The Truth from The Foul Grasp and Persecution of Superstition; And Obtain for the Island of Great Britain The Noble Appellation of The Focus of Truth; Whence Mankind Shall Be Illuminated, and the Black and Pestiferous Clouds Of Persecution and Superstition Be Banished from the Face of the Earth; As the Only Sure Prelude to Universal Peace and Harmony Among the Human Race. In Which a Sketch of a Proper System For the Education of Youth Is Submitted to Their Judgment* (London: R. Carlile, 1821), 8.

27. [T.J. Hogg], "Percy Bysshe Shelley at Oxford," *New Monthly Magazine* 34, no. 133 (January 1832): 90–96, at 95.

28. See Mark Kipperman, "Coleridge, Shelley, Davy, and Science's Millennium," *Criticism* 40, no. 3 (1998): 409–36. The debates about science in the age of *Frankenstein* were not only about transcendental versus materialist grounds but also about "the social ends of such knowledge" and "where the real power to guide and change an emerging technical-industrial era would come" (411).

29. Humphry Davy, *A Discourse, Introductory to a Course of Lectures on Chemistry Delivered in the Theatre of the Royal Institution on the 21st of January, 1802* (London: Royal Institution, 1802); reprinted in *The Collected Works of Sir Humphry Davy*, ed. John Davy, vol. 2 (London: Smith, Elder, 1839), 318; subsequent references given parenthetically.

30. Joseph Cottle, *Reminiscences of Samuel Taylor Coleridge and Robert Southey* (London: Houlston and Stoneman, 1847), 329.

31. For Davy's impact, see Christopher Lawrence, "The Power and the Glory: Humphry Davy and Romanticism," *Romanticism and the Sciences*, ed. Andrew Cunningham and Nicholas Jardine (Cambridge: Cambridge University Press, 1990), 213–27. Siobhan Watters' essay in this volume provocatively elaborates the elements of food science seeded in *Frankenstein*.

32. In 1801, Davy was appointed Professor of Chemistry at the Royal Institution.

33. For gender ideology in these discussions, with ethical consequences of this version of the separate spheres of male and female life, see Mellor, "*Frankenstein*: A Feminist Critique of Science," 287–312.

34. Humphry Davy, "Theories concerning Galvanism," in *A Syllabus of a Course of Lectures on Chemistry, Delivered at the Royal Institution of Great Britain* (London: Royal Institution, 1802), Part II, Division IV ("Of Galvanism"), Section 8, at 68.

35. So said *Quarterly Review* in July 1819 (22, no. 43), on the first page of a huge opening account of the debate and the relevant publications, 1–34: Art. I, with the running head "Abernethy, Lawrence, & c, *on the Theories of Life*." Marilyn Butler gives the author as George D'Oyley ("Frankenstein and Radical Science," 312; she misreports the date of the issue as November). Reverend D'Oyley was Christian Advocate at Cambridge University. The *Quarterly* had reviewed *Frankenstein* in March 1818 (XVIII.36, Art V, 379–84). The July 1819 article was sharply hostile to Lawrence's Frankenstein-science, indicting him for damaging himself, his profession, and the general welfare of humanity with "pernicious and degrading doctrines" advancing "opinions hostile to religion": "It appears to us imperative on those who have the superintendence of the Royal College of Surgeons, to make it an indispensable condition of the continuance of Mr. Lawrence in the office of lecturer, not only that he should strictly abstain from propagating any similar opinions in future, but that he should expunge from his lectures already published all those obnoxious passages which have given such deserved offence, and which are now circulating under the sanction of the College" (33–34). In order to stay viable, Lawrence recanted and retracted. But under the ruling of Lord Eldon in 1817, a blasphemous publication had no copyright protection, and so it was pirated and widely circulated (among the agents, Richard Carlile, who praises Lawrence in his *Address* of 1821).

36. John Abernethy, *An Enquiry into the Probability and Rationality of Mr Hunter's "Theory of Life"* (London: Longman & c, 1814), 94–95.

37. William Lawrence, *An Introduction to Comparative Anatomy, being two introductory lectures delivered at the Royal College of Surgeons, on the 21st and the 25th of March 1816* (London, 1816), 170; subsequent citations given parenthetically.

38. Erasmus Darwin, *Zoonomia; or the laws of organic life* (London: J. Johnson, 1796), Section XXXIX.4 (v. 1, p. 509).

39. For a speculative chronology for Frankenstein's career in relation to real-calendar events, see Wolfson and Levao, *The Annotated Frankenstein*, Appendix.

40. The same year Aldini published *An Account of the Late Improvements in Galvanism; with a series of curious and interesting experiments performed before the Commissioners of the French National Institute, and Repeated lately in the Anatomical Theatres of London. To which is added an appendix containing experiments on the body of a malefactor executed at Newgate* (London: Cuthell & Martin and J. Murray).

41. *London Times*, subsequently in *The European Magazine and London Review* v. 43 (1803), 74.

42. Andrew Ure, "An Account of some Experiments made on the Body of a criminal immediately after Execution, with Physiological and Practical Observations." *Quarterly Journal of Science, Literature, and the Arts* 6, no. 11(1820): 283–94, quotations at 290–91. (Read at the Glasgow Literary Society, December 10, 1818). Ure ends his essay with a prospect for bringing drowned persons back to life.

43. Thomas Beddoes, *Contributions to Physical and Medical Knowledge, principally from the West of England* (Bristol: Biggs & Cottle, 1799), a reference book for the treatment of various diseases. For the coinage, see *OED* (citing Beddoes' Introduction, 4).

44. For a brilliant discussion of language as a "godlike science," see Peter Brooks, "'Godlike Science / Unhallowed Arts': Language, Nature, and Monstrosity," in *The Endurance of Frankenstein: Essays on Mary Shelley's Novel*, ed. George Levine and U. C. Knoepflmacher (Berkeley: University of California Press, 1979), 205–20.

45. Humphry Davy, *Consolations in Travel, or the Last Days of a Philosopher* (1830), 5th ed. (London: John Murray, 1851). Davy was interested in racial strengthening by "mixture the different families of men" (41); "The Caucasian stock has always preserved its superiority,

whilst the negro or flat-nosed race has always been marked for want of intellectual power and capacity for the arts of life" (42); "The negro race has always been driven before these conquerors of the world; and the red men, the aborigines of America, are constantly diminishing in number; and it is probable that in a few centuries more their pure blood will be entirely extinct" (45). With good species husbandry, Davy envisioned a master race that would absorb the strengths and witness the extinction of "pure blood" racial inferiority.

46. On this vector, see John W. Bugg, "'Master of their language': Education and Exile in Mary Shelley's *Frankenstein*," *HLQ* 68, no. 4 (2005): 655–66; and Susan J. Wolfson. "*Frankenstein*, Race and Ethics," *Keats-Shelley Review* 34, no. 1 (2020): 12–21.

47. "In dealing with the negro, we are dealing with a being possessing the form and strength of a man, but the intellect only of a child. To turn him loose in the manhood of his physical strength . . . would be to raise up a creature resembling the splendid fiction of a recent romance." Shelley heard about the remark, recorded in *Hansard's Parliamentary Records* (*L*, 1:417–18). It was widely quoted in the United States in debates on abolition.

48. Some of my references in this paragraph and the next (including those without my own endnotes) are indebted to Theodore Ziolkowski, "Science, Frankenstein, and Myth," *Sewanee Review* 89, no. 1 (Winter 1981): 34–56, especially 53–54; Everett Mendelsohn, "'Frankenstein' at Harvard: The public politics of recombinant DNA research," *Transformation and Tradition in the Sciences*, ed. Everett Mendelsohn (Cambridge: Cambridge University Press, 1984), 317–35; and Michael Rogers, *Biohazard: The struggle to control recombinant DNA experiments, the most promising (and most threatening) scientific research ever undertaken* (New York: Random House, 1977), 196.

49. Herbert J. Muller, *The Children of Frankenstein. A Primer on Modern Technology and Human Values* (Bloomington: Indiana University Press, 1970).

50. Willard Gaylin, "We Have the Awful Knowledge to Make Exact Copies of Human Beings; The *Frankenstein* Myth is Real," *New York Times Magazine*, March 5, 1972, 12–13, 41–44, 48–49.

51. Alfred Velluci, quoted in *Time*, April 18, 1977, 45.

52. "Is Harvard the proper place for Frankenstein tinkering?," *Washington Star*, June 16, 1976.

53. Quoted in David W. Ehrenfeld, *The Arrogance of Humanism* (Oxford University Press, 1976), 96.

54. This is the chapter title in Rogers' *Biohazard*, 191–205.

55. Ted Howard and Jeremy Rifkin, *Who Should Play God? The Artificial Creation of Life and What it Means for the Future of the Human Race* (New York: Delacourte, 1977).

56. William Barrett, *The Illusion of Technique: The Search for Meaning in a Technological Civilization* (New York: Anchor, 1978), 24.

57. William Safire, Opinion columns, *New York Times*, February 27, 1997, and July 5, 2001.

58. Stephen Levick, letter to the editor, *New York Times Magazine*, April 24, 2005.

59. See Mark McCutcheon's essay in this volume for more. Just before I delivered The Charles E. Robinson Memorial Lecture at Stevens Institute of Technology on April 16, 2018 (the basis of this chapter), there was an iteration in the *New York Times*, April 13 (John Yoo and Saikrishna Prakash, "Opinion: Of Course Trump Can Fire Mueller. He Shouldn't"), on Robert Mueller's investigation of President Trump: "Ken Starr's investigation of Bill Clinton lasted for years, consumed enormous resources, and resulted in few convictions. By the end, Congress allowed the independent counsel law to quietly die. Resurrecting this Frankenstein would once again strike a blow at the separation of powers, which protects individual liberty as surely as the Bill of Rights itself." The resurrection of the trope of "Frankenstein" needed no annotation.

60. Frank Bruni, "The Internet will be the Death of Us," *New York Times*, Opinion, October 30, 2018. This peril was heralded by Kevin Roose, "Facebook's Frankenstein Moment: A Creature That Defies Technology's Safeguards," the above-the-fold headline in *New York*

Times, Business Day, September 22, 2017: "scientist Victor Frankenstein realizes that his cobbled-together creature has gone rogue" (B1).

61. Nicholas Wade, "Scientists Seek Ban on Method of Editing the Human Genome," *New York Times*, March 19, 2015.

62. Amy Harmon, "Human Gene Editing Receives Science Panel's Support," *New York Times*, February 14, 2017.

63. Chelsea Leu, "Crispr Will Change the World," *New York Times*, November 19, 2017. Science section 7–9.

64. Mike Bygrave, "False Dawns in the Brave New World of Genetics," *The Guardian*, December 22, 2002. Science section.

65. Mark Prigg, "US Government to give go-ahead to controversial chimera research to grow part-animal part-human 'Frankenstein' organs for transplants," *Daily Mail*, August 5, 2016.

CHAPTER 2

Frankenstein Meets the FAANG Five

Figures of Monstrous Technology in Digital Media Discourse

Mark A. McCutcheon

In business parlance, the "FAANG group" is shorthand for Facebook, Amazon, Apple, Netflix, and Google: the group of US-based, multinational corporations whose digital services and technologies command a considerable share of both the world economy and the attention of internet users. Over the past few years, the products and actions of these FAANG firms—for instance, social media platforms (like Facebook) and artificial intelligence or AI (one of Google's main research and development foci)—have prompted *Frankenstein*-inspired headlines and commentary in journalism, the blogosphere, and other domains of popular culture.

These popular representations of FAANG firms and products as machines or other kinds of entities run amok furnish further evidence for, and grounds for building on, the argument advanced in my recent book, *The Medium Is the Monster: Canadian Adaptations of Frankenstein and the Discourse of Technology*. That study's twofold argument holds, first, that Mary Shelley's *Frankenstein* effectively reinvented the meaning of the word "technology" for modern English and, second, that the media theory of Marshall McLuhan cemented and popularized this Frankensteinian sense of technology as human-made monstrosity, especially with reference to media technologies.

This chapter first reprises that argument and then, to illustrate and elaborate that argument, turns to some popular representations of FAANG technologies and activities and discusses the particular ways in which *Frankenstein* shapes and shadows these representations. *Frankenstein*, together with certain significant adaptations of it, has shaped and inflected the modern meaning of the word "technology" itself, such that today we can hardly speak of technology without inevitably also conjuring *Frankenstein*.[1]

To document the modernization of the meaning of the word "technology" as a discursive effect of Mary Shelley's novel, let us examine the discourse of technology in that author's day and then focus on *Frankenstein*. I retrace the early nineteenth-century redefinition of the word (from describing the study of any art to describing industrial systems) and argue that Shelley's characterization of the Creature, according to five prevailing tropes, in turn conditions the modern meaning of "technology," as some period uses of the word suggest.[2] *Frankenstein* does not explicitly use the word "technology" (just as it does not name its antagonist), but through the Creature's characterization, the novel became a literary "threshold of epistemologization,"[3] a textual battery that charged the episteme of Romantic science and culture to generate the modern discourse of technology.

In the discursive history of technology and in the literature on the relationship between *Frankenstein* and technology, we find two premises that my argument challenges. First, accounts of the provenance of technology—like that of the *Oxford English Dictionary* (*OED*)—suggest that it was in the late nineteenth century that the word attained its modern redefinition to mean, in general, tools and machines, and techniques and systems for their use, or combinations thereof. But a close reading of *Frankenstein* suggests that this modern meaning emerged much earlier in the nineteenth century—hence the second premise I question: that while *Frankenstein* is widely read as "the first and most enduring symbol of modern technology,"[4] its relationship to the discourse of technology is constructed retrospectively, as though this definitively modern discourse emerged later and could be retroactively applied to *Frankenstein*. If technology has popularized a certain interpretation of *Frankenstein*, it is because *Frankenstein* itself conditioned the modern redefinition of technology.

Prior to that period, the word "technology" had a relatively rare and specialized meaning and use in English, to refer to the study of any art or craft: "a discourse or treatise on the arts," as the *OED* offers for this antiquated definition. (And in early-modern English, "the arts" was a phrase with a more expansive meaning than it has now.) Since the early nineteenth century, the word "technology" has instead come to refer to mainly industrial machines, their operating techniques, and the systems they belong to: "machinery, equipment, . . . the mechanical arts and applied sciences," as the *OED* offers for the word's modern meaning. The *OED*, like many "keywords" essays about technology (by Raymond Williams, Andrew Ross, and Scott McQuire, among others), dates the emergence of the word's "machinery" meaning to the mid- to late nineteenth century.[5]

But I suggest that *Frankenstein* effected that modern redefinition. If, as Laura Kranzler writes, "the problematics of technological development and application are initially codified in [Mary] Shelley's work,"[6] this idea is worth taking at its word—and worth reading for the historical evidence between its lines. A close reading of *Frankenstein* shows how "technology" began circulating in its modern sense as a kind of "Frankenpheme." "Frankenpheme" is Timothy Morton's term for an "element of culture that [is] derived from *Frankenstein*, but [is] less than a work of art in completion or scale"—that is, a Frankenpheme is not so much an adaptation as a specific kind of allusion or meme.[7] In Mary Shelley's novel, we find a series of tropes that show how its language, together with its plot of uncontrolled research and monstrous result, conditions the modern discourse of technology. These tropes are utility, supplementarity, contagion, shock, and revolution.

Utility

The rhetoric of utility permeates the text. Both Victor Frankenstein and his interlocutor Walton sometimes sound like a parody of Jeremy Bentham. *Frankenstein* engages with Bentham and his philosophy of Utilitarianism in its exploitation of the then-dubious reputation of medical doctors, who traded with grave robbers to obtain cadavers at a time when Bentham worked to legitimize dissection—with legislation and, ultimately, with the donation of his own body for scientific display to promote the "further uses of the dead to the living."[8] Like Bentham, Victor pursues his research with utilitarian idealism, buoyed by "visions of extensive usefulness" (I:I.9–10).[9] But the story renders these visions ultimately ironic. Walton, for his part, first foreshadows Victor's "visions of usefulness" by imagining "the inestimable benefit which [he] shall confer on all mankind" (I:Letter I.4), and finally echoes Victor's failure, as he abandons his own "hopes of utility and glory" (III:VII.180).

Supplementarity

As the preceding passages show, Shelley applies something like Bentham's "greatest happiness principle" to the trope of utility. Walton and Victor both envision the "extension" of the "utility" of their projects for all humankind. And the modern discourse of technology is nothing if not a discourse of uncanny difference from the ontological category of the human.[10] This simultaneous opposition and intimacy between technology and humanity also informs Marshall McLuhan's famous definition of technologies as "extensions of man,"[11] extensions that sometimes act as prosthetics—and other times as replacements. "What really makes the novel . . . disturbing," writes Morton, "is not the creature's difference from, but his *similarity to* human beings."[12] Frankenstein's Creature—both human and "superhuman" (II:II.79), dead and alive—becomes a prototypical figure of technology, posing a dangerous supplement to humanity, as technology poses a dangerous supplement to capitalism: as the extension *and* replacement of human labor and agency.

Contagion

Frankenstein's images of contagion relate to its images of revolution (on which more will be said momentarily), unsurprisingly, given England's apprehension in the period over the potentially Channel-crossing infectiousness of revolutionary philosophy and foment. The passage in which Victor ponders what might come of the bride he builds for his Creature furnishes a primal scene for the modern discourse of technology: "a race of devils would be propagated upon the earth, who might make the very existence of the species of man a condition precarious and full of terror" (III:III.139). Victor's envisioned "race of devils" prefigures the way in which many technologies are represented today, from fossil fuels to the internet to nuclear energy and of course biological weaponry; in Susan J. Wolfson's chapter in this volume, she elaborates further on *Frankenstein*'s cultural role as a "cautionary tale" that shadows modern science's "dark dreams."[13]

Shock

Frankenstein dramatizes the electrifying sense of shock that has been significant both for subsequent adaptations and for modernizing the meaning and connotations of technology. While Victor's reference to "the spark of being" (I:IV.41) that animated his creation is famously ambiguous, the science of galvanism is an equally famous context for the novel, and electricity is mentioned in other scenes that inform both the method and affect of the Creature's construction. An early, foreshadowing anecdote from Victor's childhood describes "the shock" of lightning that "utterly destroyed" an oak tree and frames Victor's account of learning about "that power"—electricity and galvanism—which precipitates his own intellectual revolution, the "overthrow" of Agrippa and the alchemists (I:I.24). Shock is also a superb contranym—it means both an extremity of feeling and its nullification—and figures in McLuhan's theory of technology as a force that, in creating new environments, also shocks or numbs subjects who are accustomed to old environments.

Revolution

The instability and danger of the Creature figure the trope of revolution as another modern discursive condition for technology. As has been widely researched, the conflict between Victor and the Creature stages a drama of revolution that responds to both the French Revolution and the Luddite revolts.[14] These images of revolution speak to the spirit of Mary Shelley's age. But they also encode the motion of drastic and disruptive social change that has become integral to representations of technology: from Marx's "faith in the revolutionary potential of technology" to McLuhan's theory of technology as revolutionary change,[15] in which new media replace or consume old media and electric media produce social upheaval on a global scale.

The connected tropes of utility, supplementarity, revolution, contagion, and shock converge significantly in the repeated claim, made first by the dying Victor and then by his Creature, that the latter is an "instrument of future mischief,"[16] anticipating the popular sense of technology as an instrument of "future shock" (as in the title of Alvin Toffler's popular 1970 book). While *Frankenstein* leaves technology, like its antagonist, unnamed, it supplies a primal scene for redefining technology—not after midcentury but as early as the 1820s.

Three representative articulations of the modern discourse of technology, in the 1820s and 1830s, point to its Frankensteinian conditioning, as evoked and evinced in the writings of the aspiring auto-icon Bentham, the Harvard professor Jacob Bigelow, and the steam-power advocate Thomas Love Peacock. Bentham's 1827 *Rationale of Judicial Evidence* refers to "questions in technology" in a way that can be read to use the word in either its premodern or its emerging modern sense, especially in the way Bentham distinguishes the term from "science," as the terms are now conventionally distinguished in current English.[17] Then, in 1831, the year Mary Shelley's revised *Frankenstein* was published, so was Bigelow's book *Elements of Technology*, which—in suggestively Frankensteinian language—explicitly "undertakes" a "revival" of the word "technology" to "embody" the industrial and practical arts collectively.[18] Lastly, Peacock, an acquaintance of the Shelleys and Bentham, in two of his novels uses the word in ways that articulate modernity and menace, bridging the premodern and modern senses of "technology." *Crotchet Castle*, for instance, refers to political economy as "a hyper-barbarous technology,"[19] evoking both the older sense of the word as a knowledge system and its newer, Frankensteinian sense: as a kind of violence.

If Mary Shelley's novel shapes the modern meaning of "technology" in its distinctive characterization of a manufactured monster according to the aforementioned tropes, then McLuhan's writings consistently bring the figure of that monster to bear on his characterizations of technology and media.[20] The title of McLuhan's first book, *The Mechanical Bride*, allude to the bride of Frankenstein; the book's title essay explicitly cites "Frankenstein fantasies" of "the horror of a synthetic robot running amok" to explain the alienating effects of commercialism and automation in a cultural landscape dominated by imagery of "sex, technology, and death."[21] And McLuhan's landmark study, *Understanding*

Media, opens with an ominous vision of that quintessentially Frankensteinian technology, artificial intelligence: "Rapidly, we approach the final phase of the extensions of man—the technological simulation of consciousness."[22]

McLuhan strove throughout his career to understand media by adopting a neutral stance of critical detachment. But this detached stance proved both controversial and contradictory: controversial, in that his refrain from value judgments led to the widespread misreading of him as an antibook techno-fetishist; and contradictory, in that his own writings—in tone and imagery—consistently reveal his deep hostility to technological change. McLuhan amplifies the tropes of utility, supplement, revolution, shock, and contagion with which *Frankenstein* modernized the meaning of "technology," and his use of the word as a Frankenpheme both undergirds his deterministic premise concerning technology and undermines his declared suspension of judgment. For McLuhan, new technologies produce pain, confusion, and despair,[23] to which individuals and societies respond by going into a kind of shock or "autoamputation": "With the arrival of electric technology, man extended, or set outside himself, a live model of the central nervous system itself, . . . a development that suggests a desperate and suicidal autoamputation."[24] McLuhan consistently describes technology in terms of invasion,[25] disease,[26] disaster,[27] and conflict on a global scale: "every new technology," he said grimly in 1968, "necessitates a new war."[28] All five technological tropes that we find in *Frankenstein* thus inform and shape McLuhan's model of global technological change, a model figured in Frankensteinian imagery that clearly conveys his hostility to such change.

An uncommonly clear public expression of McLuhan's Frankenpheme of technology—and a rare public confession of his hostility to it—was printed in a 1969 issue of *Playboy*. Pressed by the interviewer to clarify his opinion about "new technology" as a "revolutionizing agent," McLuhan replied frankly, "I view such upheavals with total personal dislike and dissatisfaction. . . . I derive no joy from observing the traumatic effects of media on man, although I do obtain satisfaction from grasping their modes of operation. . . . It's vital to adopt a posture of arrogant superiority; instead of scurrying into a corner and wailing about what media are doing to us, one should charge straight ahead and kick them in the electrodes. They respond beautifully to such treatment and soon become servants rather than masters."[29] In his *Playboy* comments, McLuhan figures technology as a rebellious, male artificial intelligence, whose inherently rebellious tendency should be violently preempted; technology is a menace that needs to be subordinated—mastered—to be useful. This figure vividly condenses and dramatizes many of the tropes with which Mary Shelley characterizes the Creature.

As a maverick public intellectual embraced by the 1960s counterculture and Madison Avenue alike, McLuhan popularized a Frankensteinian sense of technology that Shelley's *Frankenstein* had prototyped. It is telling that both their work has been adapted together in more recent popular and scholarly texts concerning technology, some of the best known and most influential of which include William Gibson's 1984 novel *Neuromancer*, David Cronenberg's 1983 film *Videodrome*, and Friedrich Kittler's monograph *Gramophone Film Typewriter* (first published in 1986 and translated into English in 1999). If *Frankenstein* shaped the modern meaning of "technology" as manufactured monstrosity, then McLuhan's Frankenpheme of technology and subsequent adaptations of both these writers' works have globally popularized this meaning. Since the publication of *Frankenstein*, the word "technology" has become valorized as a policy and business imperative, but it has also become popularly evoked as a Frankensteinian image of manufactured monstrosity, a discursive instrument of future mischief.

Several contemporary phenomena exemplify this ambivalent evocation of technology. Glaring examples include two human-made projects of modernity that have run amok and now threaten existential ecological catastrophe: nuclear weaponry[30] and fossil-fuel-based industry and infrastructure.[31] Less obvious but no less globally (albeit differently) implicated (if differently so) examples include digital communications, especially social media, and the corporations that have innovated them and popularized their use: the so-called FAANG companies. In the remainder of this chapter, I consider three examples in particular: the controversy over Facebook's and Twitter's impact on US political governance; Apple's 2016 advertisement "Frankie's Holiday"; and Amazon's acquisition and redistribution of the TV series *The Expanse*.

Since the nomination of Donald Trump to the US Republican Party leadership and his subsequent election as US president, numerous commentators have likened Trump to Frankenstein's Creature (as Wolfson also notes in her chapter in this book). Former US senator Harry Reid called Trump the Republican Party's Frankenstein monster on several occasions.[32] The TV celebrity pundit Bill Maher applied this analogy to Trump as early as 2015,[33] and the *Washington Post* has repeatedly applied it too.[34] In the run-up to the November 2016 US presidential election, former president Barack Obama did not explicitly make such a comparison, but he did describe the emergence of Trump as a viable candidate in terms that explain why the *Frankenstein* allusion has become so prevalent in this context: "The problem is not that all Republicans think the way this guy does. The problem is, is that they've been riding this tiger for a long time. . . . So the point is, if your only agenda is either negative—negative is a euphemism, crazy—based on lies, based on hoaxes, this is the nominee you get. *You make him possible*."[35]

The *Frankenstein* allusions continued during Trump's time in office.[36] An article in the *Atlantic*, speculating on the possible destruction of the position of the president, describes Trump as "a Frankenstein's monster of past presidents' worst attributes."[37] And it was during Trump's presidential term that the *Frankenstein* analogy took on a specifically technological character, amid the controversy that emerged over the unexpectedly significant role that social media played in influencing the outcome of the 2016 US presidential election.

A 2018 *Fast Company* article about Twitter details the problems and pitfalls facing that firm, particularly in connection with the Trump administration. Identifying "issues that have long bedeviled Twitter," like "abuse, the weaponizing of anonymity, bot wars, and slow-motion decision making by the people running a real-time platform," the journalists write that "these problems have only intensified since Donald Trump became president and chose Twitter as his primary mouthpiece."[38] The article quotes several Twitter employees who make the *Frankenstein* connection between their platform and the Trump administration. One source does so with reference to one of *Frankenstein*'s own intertextual sources, the classical myth of Pandora: "Safety got away from Twitter," says a former vice president at the company. "It was Pandora's box. Once it's opened, how do you put it all back in again?"[39] Another Twitter employee makes the comparison somewhat more explicitly, here in terms of the monstrous: "On November 8, employees were shocked as the election returns poured in, and the morning after Trump's victory, Twitter's headquarters were a ghost town. Employees had finally begun to take stock of the role their platform had played not only in Trump's rise but in the polarization and radicalization of discourse. 'We all had this "holy shit" moment,' says a product team leader at the time, adding that everyone was asking the same question: 'Did we create this monster?'"[40]

More openly still does Kevin Roose, in a 2017 article for the *New York Times*' Technology section, liken social media to *Frankenstein* in connection with the 2016 presidential election. His article, titled "Facebook's Frankenstein Moment," features a still frame from the creation scene in Hammer Films' 1957 *Curse of Frankenstein* and frames its account of Facebook's "series of scandals" as an adaptation of *Frankenstein*'s core plot of a manufactured monster that turns on its maker and runs amok: "If I were a Facebook executive, I might feel a Frankensteinian sense of unease these days. The company has been hit with a series of scandals that have bruised its image, enraged its critics and opened up the possibility that in its quest for global dominance, Facebook may have created something it can't fully control."[41]

Roose's article emphasizes the Frankensteinian sense of technology not only because of the section of the paper it appears in but also in its language: "Facebook was simply not built to handle problems of this magnitude. It's a technology company . . . in the business of building apps and selling advertising, not determining what constitutes hate speech." One of Roose's sources, the author Antonio García Martínez, underscores this sense, referring to Facebook as an unpredictable "machine" and explaining that Facebook's corporate leaders "still see themselves as a technology middleman." And the article closes by putting Facebook's technological implications in a global context that resonates grimly on several levels: "The company can't dodge responsibility for the world it has helped to build. In the future, blaming the monster won't be enough."[42]

Another of the FAANG firms, Apple, released a television advertisement in late 2016 that further cements the modern sense of "technology" as manufactured monstrosity—while also implicitly critiquing the cultural and political culture that helped Trump win the US presidency earlier that same autumn. "Frankie's Holiday," directed by Lance Acord, stars Brad Garrett as the Creature, who in this story leaves his cozy cottage home in the mountains to try caroling in the nearby village square. Based on Boris Karloff's iconic portrayal in James Whale's 1931 film *Frankenstein*, this Creature also has neck bolts, but he replaces them with festive Christmas lights. His caroling effort draws a crowd of onlookers, who seem hostile until a little girl fixes one of his neck-bolt lights and starts singing along with him, whereafter the crowd joins in. The ad's key dramatic device—the Creature's substitution of holiday lightbulbs for bolts—emphasizes the creature's figuration of technology, specifically electric technology, with its aforementioned evocations of shock, an affect performed in the ad spot by the mostly silent, still crowd of onlookers.[43]

Additionally, the advertisement's pointed focus on *Frankenstein*'s theme of character's social construction led several commentators to read it as an allusive response to a presidential victory resulting from the exploitation of atavistic fear and hatred.[44] "The company's ad is intended to deliver a message of unity at a time when the U.S. and the world are experiencing disunity," wrote Don Reisinger in *Fortune* magazine.[45] "Frankie's Holiday" thus anticipates and dramatically encapsulates the close coordination of technological and sociocultural concerns in journalistic and popular cultural representations of both US politics and FAANG economics.

In the context of FAANG economics especially, Amazon's recent acquisition, redistribution, and continuance of production of the science fiction TV series *The Expanse* presents a richly complex digital example of the Frankenpheme of technology. Based on James S. A. Corey's best-selling series of science fiction novels, *The Expanse* posits an interplanetary postcolonial society, set several centuries in the future, in which humankind has colonized the solar system and reorganized into three main territorial groups: Earthers, Martians, and "Belters," the latter being the largely working-class citizens of the asteroid belt. The series's plot in the first three seasons mainly concerns a "first contact" type of encounter with extraterrestrial technology, which—by underregulated, corporate-sector R&D—becomes weaponized, galvanizing an interplanetary arms race and, thus, in fine Frankensteinian fashion, poses an existential threat to the whole solar system. Echoing McLuhan's aforementioned claim about technology and war, a Martian ambassador rhetorically asks his Earth counterpart an eminently Frankensteinian question: "Why is all new technology first viewed as a weapon?"[46] The series openly acknowledges its debt to *Frankenstein* in a scene where two protagonists engage in a little literary criticism, as one removes the stitches from the other's head injury:

> AMOS: I don't want you looking like Frankenstein when Mei sees you.
> PRAX: Frankenstein was the name of the doctor. The monster didn't have a name.
> AMOS: God damn—that's right.[47]

Other Frankensteinian details include the spectral return of a police detective who haunts the protagonist, Jim Holden, as a kind of "simulation":[48] "Miller's ghost was an artifact of the alien technology that had created the gates and a dead man."[49] The show and its source novels thus develop a distinctly Gothic tone that is unusual for the space opera genre. And much like *Frankenstein*, too, *The Expanse* holds complexities, contradictions, and ambiguities. The show is intrinsically hopeful for today's zeitgeist insofar as it posits any future for humankind whatsoever; the cruelty lacing this optimism, though, lies in the series's projection of human civilization's interplanetary expansion at a time when human civilization is bracing instead for contraction, if not destruction (intimations of today's climate crisis appear during the title sequence, which shows brief scenes of rising ocean levels, and in exterior shots like those showing a "Yukon archipelago"); indeed, at the time of this writing, societies worldwide are actively contracting (via border lockdowns, business closures, and distancing measures) in order to mitigate the spread and impact of the COVID-19 pandemic. Similarly, the structural staples that build and sustain *The Expanse*'s hopeful future include fantastical technologies (e.g., rockets and engines that turn interplanetary travels into round trips, not one-way missions), but they also include real-world systems and institutions—not only corporate business, which is consistently characterized as sociopathic in characters such as Dresden and Murtry, but also, more intriguingly, its opposite: organized labor, which is often depicted positively and thus adds further subtle but significant Frankensteinian dimension to the show.

An important institution in *The Expanse* is the Outer Planetary Alliance, or OPA: a networked collective operating in the Belt and farther-flung gas giant moons, the OPA "had begun its life more like a labor union than a nation," Corey writes;[50] as the series progresses, the OPA becomes a major political power in its own right. Several protagonists are current or former OPA members, and the series's plots often take shape around the kinds of radical democratic practices with which unions organize themselves and which they espouse and spread to other domains of social life: consensus-based decision-making, trust in expertise, bargaining, the nurturance of skilled labor, deescalation tactics, critical reflection, and transparent communication, including the speaking of truth to power.

In *The Expanse*'s settings, characters, and plots, it thus features some of the most refreshingly positive representations of organized labor—not just in science fiction but on television generally. Union offices, representatives, and principles permeate the everyday life of *The Expanse*'s fictional world. "Bargaining is how civilizations are built," says Ashford, captain of a Belt flagship.[51] Far from simplistic, however, these representations become significantly complicated by conflicting perceptions and depictions of the OPA. OPA leader Fred Johnson asserts at one point that "the OPA is only interested in human rights and jobs for all Belters."[52] Some characters support the OPA for furthering workers' interests, while others think the OPA has sold out in bidding for political legitimacy, and still others see the OPA as a terrorist network. The resulting ambivalence over the OPA (visualized in the organization's logo, with the *A* rendered like the graffito for "Anarchy") thus furnishes additional Frankensteinian texture to the series's premise and plot, since this ambivalent detail reanimates—albeit with a critical, pointedly progressive edge—the long-standing identification of the working class with Frankenstein's composite, alienated Creature.[53] As my colleague in labor studies Bob Barnetson and I argue elsewhere, *Frankenstein*'s Creature has furnished elites with a gruesome figure for caricaturing the working class—as collective, interchangeable, and quasi-mechanical—and science fiction has long adapted this specific caricature to characterize monstrous antagonists: "Shelley's politically resonant 'hideous progeny' thus founded the trope of the collective antagonist, seen throughout SF, from the Martians of H. G. Wells' 1897 *War of the Worlds* to the Borg of *Star Trek: The Next Generation* (1987–94) and the Cylons of *Battlestar Galactica* (2004–09)."[54]

I have detailed these Frankensteinian elements of *The Expanse*'s narrative—its Frankensteinian plot of corporate hubris and catastrophic backfire, its premise in generally valorized working-class and union culture—in order to illustrate the irony arising between the series's content and its ownership. Amazon Prime Video's acquisition and redistribution of the series answered a fan-led campaign to #SaveTheExpanse, which had been canceled by SyFy. The fact that Amazon has acquired such a subversively prolabor series as *The Expanse* seems both richly ironic and yet totally typical of capitalist postmodernity. Amazon's CEO, Jeff Bezos, is currently the world's richest man,[55] and his company has gained widespread notoriety for its extensive exploitation of workers,[56] in tandem with its advances in automation and artificial intelligence (AI).

Despite this dramatic change of ownership, the show's fourth and fifth seasons have stayed true to form while deepening and nuancing its sympathetic representations of labor by expanding on related contexts of higher education and anticolonialism. By the time this chapter appears, the sixth and final season will have aired, and the screen adaptation of the sixth novel, *Babylon's Ashes*, under Amazon's watch (i.e., its pervasive use of AI in surveilling and serving Prime audiences),[57] is of particular interest, since this novel especially integrates union institutions and activities (not strikes) into its plot.

Further contradiction between ownership and content arises for the whole series, books and shows alike, insofar as *The Expanse*'s plots often hinge on the openness of access to critical knowledge and information. The protagonist, Jim Holden, embodies an ethos strongly reminiscent of Open Access and "information-wants-to-be-free" hackerdom: "Give the people the information they need. Trust them to do the right thing."[58] This ethos sometimes works and sometimes backfires, and it contrasts starkly to the proprietary distribution and accessibility of *Expanse* texts: the shows are only available to Amazon Prime viewers, and the novellas by Corey that accompany the print novels are only available as Amazon Kindle e-books. Given Amazon's category-killing global stature, demonstrably exploitative labor practices, and—last but far from least—its use of AI in monitoring and stimulating Prime Video usage, the company's adoption, production, and distribution of such a labor-friendly franchise as *The Expanse* represents a site of fascinating, Frankensteinian ironies and contradictions that are symptomatic of late global capitalism in general and FAANG business specifically.

As the latest innovation in automation, AI has become a frequent target of *Frankenstein* allusions and memes.[59] And AI has become integral to how all the FAANG firms do business, with results from the ridiculous (like when "Twitter taught Microsoft's AI chatbot to be a racist")[60] to the existentially terrifying, which brings us back to the forty-fifth US president and the demonstrated power of digital media to wreak geopolitical havoc unanticipated by their designers, to exacerbate disinformation campaigns—and thereby to inflame the global climate crisis by emboldening denialists of climate science, from said president to his format-rights counterparts in Canada, the United Kingdom, and elsewhere. The elite protectors and beneficiaries of what John Bacher incisively terms "petrotyranny"—"the deadly trinity of oil, war and dictatorship [that] presents the greatest challenge to humanity at the start of the new millennium"[61]—have entrenched a world-system of fossil-fuel-based corporate capital that now threatens Frankensteinian destruction on a horrific, planetary scale. Digital communication technologies and social media in particular thus join the ranks of technological developments—nuclear power, greenhouse gas emissions, terminator crop seeds—that all share the dubious distinction of drawing unfavorable comparisons to a human-made monster that was first imagined as an assemblage of human and animal parts that had been dead but are made to live again, only to be shunned by society and seek vengeance on it. Nevertheless, for all the Frankensteinian, technological chaos that digital media have visited, however inadvertently, on today's political and social life, as well as life per se—overcoming this ominous "Frankenstein barrier"[62] that shadows the present's possible futures can still involve time-tested, low-tech strategies like those of yesterday's Luddites or today's unions and activists. Get out the vote. Eschew air travel. Plant trees. If human hands can make a monster, they can unmake one too.

NOTES

1. Mark A. McCutcheon, *The Medium Is the Monster: Canadian Adaptations of Frankenstein and the Discourse of Technology* (Edmonton, AB: Athabasca University Press, 2018), 3, 7.

2. The section of this chapter that analyzes Mary Shelley's characterization of the Creature and period usage of the word "technology" adapts a portion of chapter 2 in *The Medium Is the Monster*; see pages 64–74 in that volume.

3. Michel Foucault, *The Archaeology of Knowledge and the Discourse on Language* (1969), trans. A. M. Sheridan Smith (New York: Pantheon, 1972), 187.

4. Martin Tropp, *Mary Shelley's Monster: The Story of Frankenstein* (Boston: Houghton Mifflin, 1976), 66.

5. Andrew Ross, "Technology," in *New Keywords: A Revised Vocabulary of Culture and Society*, ed. Tony Bennett, Lawrence Grossberg, and Meaghan Morris (Oxford, UK: Blackwell, 2005), 342–344.

6. Laura Kranzler, "Frankenstein and the Technological Future," *Foundation*, no. 44 (1988): 43.

7. Timothy Morton, *Mary Shelley's Frankenstein: A Sourcebook* (London: Routledge, 2002), 47.

8. Jeremy Bentham, quoted in Morton, 86.

9. Quotations follow the publication of 1818, as presented in *Frankenstein: Annotated for Scientists, Engineers, and Creators of All Kinds*, ed. David H. Guston, Ed Finn, and Jason Scott Robert (Cambridge, MA: MIT Press, 2017).

10. Scott McQuire, "Technology," *Theory, Culture and Society* 23, nos. 2–3 (2006): 255.

11. Marshall McLuhan, *Understanding Media: The Extensions of Man* (1964; repr., Corte Madera, CA: Gingko, 2003).

12. Morton, *Mary Shelley's Frankenstein*, 46.

13. See chapter 1 in this volume.

14. Paul O'Flinn, "Production and Reproduction: The Case of *Frankenstein*," *Literature and History* 9, no. 2 (1986): 200.

15. Ross, "Technology," 343.

16. See both 3:7.182 and 3:7.186.

17. Jeremy Bentham, *Rationale of Judicial Evidence, Specially Applied to English Practice*, vol. 1 (London: Hunt and Clarke, 1827), 19.

18. Jacob Bigelow, *Elements of Technology, Taken Chiefly from a Course of Lectures Delivered at Cambridge, on the Application of the Sciences to the Useful Arts*, 2nd ed. (Boston: Hilliard, Gray, Little, and Wilkins, 1831), iv.

19. Thomas Love Peacock, *Crotchet Castle* (1831; repr., London: H. Hamilton, 1947), 110. Citation refers to the H. Hamilton edition.

20. This and the subsequent five paragraphs summarize key highlights from the close reading of McLuhan's oeuvre elaborated in chapter 4 in McCutcheon, *Medium Is the Monster*, 85–101.

21. Marshall McLuhan, *The Mechanical Bride: Folklore of Industrial Man* (London: Routledge and Kegan Paul, 1951), 100.

22. McLuhan, *Understanding Media*, 5.

23. Marshall McLuhan and Quentin Fiore, *The Medium Is the Massage: An Inventory of Effects* (1967; repr., Corte Madera, CA: Gingko, 2001), 8.

24. McLuhan, *Understanding Media*, 65.

25. McLuhan, 30.

26. Marshall McLuhan, *The Gutenberg Galaxy: The Making of Typographic Man* (1962; repr., New York: Signet, 1969), 17.

27. McLuhan, 302.

28. Marshall McLuhan and Quentin Fiore, *War and Peace in the Global Village* (New York: Bantam, 1968), 98.

29. Marshall McLuhan, "The *Playboy* Interview," *Playboy*, March 1969, 158.

30. Morton, *Mary Shelley's Frankenstein*, 56.

31. McCutcheon, *Medium Is the Monster*, 176.

32. Tom Kludt, "Harry Reid: Trump Is GOP's 'Frankenstein Monster,'" CNN, September 29, 2016, http://www.cnn.com/2016/09/29/politics/.

33. Kellan Howell, "Bill Maher: Donald Trump Is a 'Frankenstein Monster' Created by the Tea Party," *Washington Times*, June 27, 2015, https://www.washingtontimes.com/news/2015/jun/27/.

34. For instance, see Robert Kagan, "Trump Is the GOP's Frankenstein Monster. Now He's Strong Enough to Destroy the Party," *Washington Post*, February 25, 2016, http://www.washingtonpost.com/opinions/2016/02/25; Dana Milbank, "Donald Trump Is the Monster the GOP Created," *Washington Post*, July 7, 2015, https://www.washingtonpost.com/opinions/2015/07/07; and Greg Sargent, "The Donald Trump Monster Is Unstoppable and Is Now Feeding on Itself," *Washington Post*, September 10, 2015, https://www.washingtonpost.com/blogs/plum-line/wp/2015/09/10/.

35. Juliet Eilperin, "In Ohio, Obama Blasts GOP: 'They've Been Feeding Their Base All Kinds of Crazy for Years,'" *Washington Post*, October 13, 2016, https://www.washingtonpost.com/news/post-politics/wp/2016/10/13/.

36. Steve Benen, "Dr. Frankenstein Didn't Like His Creation, Either," *Rachel Maddow Show*, MSNBC, August 23, 2017, http://www.msnbc.com/rachel-maddow-show/.

37. Jack Goldsmith, "Will Donald Trump Destroy the Presidency?," *Atlantic*, October 2017, https://www.theatlantic.com/magazine/archive/2017/10/.

38. Austin Carr and Harry McCracken, "'Did We Create This Monster?': How Twitter Turned Toxic," *Fast Company*, April 4, 2018, https://www.fastcompany.com/40547818/.

39. Carr and McCracken.

40. Carr and McCracken.

41. Kevin Roose, "Facebook's Frankenstein Moment," *New York Times*, September 21, 2017, https://www.nytimes.com/2017/09/21/technology/.

42. Roose.

43. Apple, "Frankie's Holiday," directed by Lance Acord, featuring Brad Garrett, November 2016; available on YouTube, November 23, 2016, https://www.youtube.com/watch?v=bFf3r-pT_WA.

44. Ed Mazza, "Apple's Touching New 'Frankenstein' Christmas Ad Will Bring Tears to Your Eyes," *Huffington Post*, November 21, 2016, https://www.huffingtonpost.ca/entry/apple-frankenstein-christmas-commercial_us_5833b7cbe4b030997bc10ff9.

45. Don Reisinger, "Watch Apple's 2016 Holiday Ad in Response to U.S. Disunity," *Fortune*, November 21, 2016, http://fortune.com/2016/11/21/apple-2016-holiday-ad/.

46. . *The Expanse*, season 2, episode 12, "The Monster and the Rocket," directed by Robert Lieberman, written by Mark Fergus and Hawk Ostby, featuring Jeff Seymour, SyFy, April 12, 2017; available on Amazon Prime Video, 2019, https://www.primevideo.com/.

47. *The Expanse*, season 3, episode 5, "Triple Point," directed by Jeff Woolnough, written by Georgia Lee and Hallie Lambert, featuring Wes Chatham and Terry Chen, SyFy, May 9, 2018; available on Amazon Prime Video, 2019, https://www.primevideo.com/.

48. James S. A. Corey, *Abaddon's Gate* (New York: Orbit, 2013), 264.

49. James S. A. Corey, *Cibola Burn* (New York: Orbit, 2014), 41.

50. Corey, *Abaddon's Gate*, 183.

51. *The Expanse*, season 3, episode 12, "Congregation," directed by Jennifer Phang, written by Daniel Abraham, Ty Franck, and Hallie Lambert, featuring David Strathairn, SyFy, June 27, 2018; available on Amazon Prime Video, 2019, https://www.primevideo.com/.

52. *The Expanse*, season 1, episode 4, "CQB," directed by Jeff Woolnough, written by Naren Shankar and Georgia Lee, featuring Chad L. Coleman, SyFy, December 29, 2015; available on Amazon Prime Video, 2019, https://www.primevideo.com/.

53. David McNally, *Monsters of the Market: Zombies, Vampires, and Global Capitalism* (Chicago: Haymarket Books, 2011), 88.

54. Mark A. McCutcheon and Bob Barnetson, "Resistance Is Futile: On the Underrepresentation of Unions in Science Fiction," *TOPIA*, no. 36 (2016): 158.

55. Luisa Kroll and Kerry A. Dolan, "Billionaires: The Richest People in the World," *Forbes*, March 5, 2019, https://www.forbes.com/billionaires/#78ac6586251c.

56. Shannon Liao, "Amazon Warehouse Workers Skip Bathroom Breaks to Keep Their Jobs, Says Report," *Verge*, April 16, 2018, https://www.theverge.com/2018/4/16/.

57. Blake Morgan, "How Amazon Has Reorganized around Artificial Intelligence and Machine Learning," *Forbes*, July 16, 2018, https://www.forbes.com/sites/blakemorgan/2018/07/16/.

58. Corey, *Abaddon's Gate*, 521.

59. For instance, see Amanda Davis, "What 'Frankenstein' Can Teach Us about the Future of AI and Robotics," The Institute, *IEEE Spectrum*, May 15, 2018, http://theinstitute.ieee.org/tech-history/technology-history/what-frankenstein-can-teach-us-about-the-future-of-ai-androbotics; and Catherine Stinson, "Deep Learning: Why It's Time for AI to Get Philosophical," *Toronto Globe and Mail*, March 25, 2018, https://www.theglobeandmail.com/opinion/.

60. James Vincent, "Twitter Taught Microsoft's AI Chatbot to Be a Racist Asshole in Less than a Day," *Verge*, March 24, 2016, https://www.theverge.com/2016/3/24/.

61. John Bacher, *Petrotyranny* (Toronto: Dundurn, 2000), 17.

62. George Slusser, "The Frankenstein Barrier," in *Fiction 2000: Cyberpunk and the Future of Narrative*, ed. George Slusser and Tom Shippey (Athens: University of Georgia Press, 1992), 46–71.

CHAPTER 3

"THE HISTORY OF GODS"

SINGULARITY AND GENDER IN *EX MACHINA*

Lisa Crafton

In 2015, Ashok Goel, a computer science professor at Georgia Tech who was grappling with over four hundred online students in a class on artificial intelligence, created Jill Watson, an AI teaching assistant based on IBM's Watson platform. Jill Watson is perhaps best known as the computer that beat two *Jeopardy* champions. In a TED Talk that has been seen by over thirty thousand viewers, Goel says that Watson's creation involved quite a bit of trial and error: Watson went through several prototypes, which was, Goel said, something like "raising a child." The students in that AI class were unknowingly interacting online with the artificial TA, who answered online questions adeptly and quickly as soon as they were posted. "Her" teaching evaluations were some of the highest ever recorded at Georgia Tech, with one student even nominating Watson for Outstanding Teaching Assistant. Once the professor informed the students about Watson's identity as an AI, student responses were all enthusiastically supportive of the experiment; one even asked if Watson could "come out and play."[1]

This kind of technophilia—our seduction by the technology we create—is addressed in sobering ways by contemporary critics who raise concerns about the existential stakes and the social consequences of an unchecked teleological march toward technological progress. A significant contribution to this ongoing, critical conversation is the recent annotated version of Mary Shelley's *Frankenstein* published in celebration of the bicentennial of the novel. The 2017 MIT publication bears the subtitle *Annotated for Scientists, Engineers, and Creators of All Kinds* and pairs the 1818 version of the manuscript—line edited and amended by the masterful Charles E. Robinson, to whose memory this volume of essays is dedicated—with annotations and essays by leading scholars exploring the social and ethical aspects of scientific creativity. The text, in fact, links passages from Shelley's novel to commentary from scholars on such issues as synthetic biology, artificial intelligence, and robotics. One of the most interesting annotations is keyed to a critical moment in the text where Victor acknowledges both his remorse over making the Creature and his complicity in the Creature's horrendous acts. After Victor has watched Justine, the trusted servant in the Frankenstein household, be accused of and executed for the murder of his younger brother, William, which the Creature committed, Victor acknowledges his own guilt: "I had been the author of unalterable evils; and I lived in daily fear, lest the monster whom I had created should perpetrate some new wickedness" (II.I.73).[2] Sheldon Krimsky offers this annotation to the episode:

> The remorse Victor expresses is reminiscent of J. Robert Oppenheimer's sentiments when he witnessed the unstable power of the atomic bomb. A passage from the Hindu scripture of the Bhagavad-Gita flashed before Oppenheimer's mind: "I am become death, the destroyer of worlds." In this short phrase, Oppenheimer, as one of the architects of the A-bomb, acknowledged that he had unleashed a force that could lead to the annihilation of civilization. . . . Victor's responsibility for his horrific scientific experiment has already passed. It appears that the creature is beyond control. . . . Oppenheimer, who witnessed a test of the atomic bomb at Los Alamos in 1945, still had an opportunity to prevent the use of the bomb against humans. . . . Those scientists with high moral conscience view their responsibility to warn about the malevolent uses of scientific results to their students, to their colleagues, and to the public. They would cease and desist from scientific research that has no redeeming value but destruction or baneful dehumanization. Victor's anguish is a warning to those scientists who bracket away the moral quality of their work under a banner of pure inquiry, whatever its outcome. Whether it is cloning a human being, creating a new biological weapon, releasing transgenic species, or designing human genomes, these ends call out for acts and acknowledgments of social responsibility. (II:I.73n4)

The MIT annotated *Frankenstein* makes palpably evident the integral function that Shelley's text serves for a generation's ethical quandaries about emerging science or technology.

No genre has more widely catalogued these anxieties than film. Caroline Picart has used the term "Frankenstein cinemyth" to denote the diverse appropriations of the Frankenstein mythos in film, each adaptation representing the cultural anxieties of its moment (eugenics, race, technology, etc.). Discussing multiple filmic adaptations across genres, Picart distinguishes between those adaptations that provide a closure that destroys the Creature and suggests a return to "normal" (and significantly heteronormative normal, as in Whale's 1931 film *Frankenstein*) versus those contemporary, looser "adaptations" that unleash more problematic issues and provide unstable if not unhappy closure.[3] In the first quarter of the twenty-first century, there is no better filmic example of this kind of dystopian reworking of Shelley's tale than *Ex Machina*.[4] Released in 2015, Alex Garland's film adapts the Frankenstein myth within the realm of AI, destabilizing categories of human and machine and interrogating notions of "singularity," especially through gender, thus offering a contemporary inquiry into the "technologies" of Frankenstein.[5]

Time magazine's 2011 cover read, "2045: The Year Man Becomes Immortal," and the eponymously titled article summarized decades of discussion among futurists, scientists, and science fiction writers on the question of singularity—the hypothesis that the invention of artificial intelligence will abruptly trigger runaway technological growth, resulting in unfathomable changes to human civilization. Almost two hundred years before this provocative magazine cover appeared, Shelley wrote of the Promethean inventor Victor Frankenstein and his Arctic-explorer foil, Captain Walton, both embracing such a moment of singularity, although they do not use that term. Walton boasts of "the inestimable benefit which [he] shall confer on all mankind, to the last generation" (I:Letter I.4), and Victor's experiment begins at least with such idealistic goals: "Life and death appeared to me ideal bounds, which I should first break through, and pour a torrent of light into our dark world. A new species would bless me as its creator and source" (I:III.37). These positive visions, that human ingenuity could bring about a kind of technotopia, do not last for long in the novel, and contemporary notions of singularity have been pessimistic. Vernor Vinge, a computer scientist and science fiction writer, asserts in his 1993 *The Coming Technological Singularity* that the new superintelligence will continue to upgrade itself and advance technologically at an incomprehensible rate, signaling the end of the human era.[6] It is this moment into which the audience is thrust in Garland's film *Ex-Machina*.

As Garland recounts his writing and submission of the script, he says he felt "pleased when [he] hit upon this plot" but became "dismayed" when he discovered that a spate of other AI movies had been made in the few years prior: "There was *Her* (2013) which tracked a love affair between a man and his computer's sentient operating system. Next was *Transcendence*, in which Johnny Depp uploaded his mind to the Internet. In *Automata* and *Chappie*, servile robots acquired independence. In *Big Hero 6*, an AI befriends a boy. And in *Avengers: Age of Ultron* . . . a robot tries what we all know they're secretly planning: the destruction of mankind. Among filmmakers, there was an AI party going on, to which we were late."[7] The film's title, *Ex Machina*, can translate to "out of the machine" and invokes, of course, the Latin phrase "deus ex machina," which means "a God from/out of a machine." In its original context, the phrase refers to the crane that held a god over the stage in ancient Greek and Roman drama. The practice of introducing a god at the end of a play to unravel and resolve the plot dates from at least the fifth century BC. Since the late 1600s, "deus ex machina" has been applied in English to unlikely saviors, and the term has evolved to mean a plot device whereby a seemingly unsolvable problem is suddenly and abruptly resolved by the inspired and unexpected intervention of some new event. Garland's title significantly, and perhaps ominously, omits the "deus," leaving us only the "machina." The film's closure deploys an image of a kind of deus ex machina, as we will see, but the title's erasure of the god is intentionally and thematically provocative.

The film's scientist figure, Nathan, the creator of AI, assumes godlike status in a manner that seems true to the *Frankenstein* novel as well as the *Frankenstein* cinemyth. Nathan, a Silicon Valley Victor Frankenstein, is a computer mogul responsible for creating the platform Blue Book (a fictional version of Google) and the creator of female AIs. He chooses the computer geek Caleb to participate in his experiments as the human component in the Turing test on the recent AI model, Ava. Welcoming Caleb to his eclectic compound, what Manohla Dargis calls a "modernist retreat that is part Zen palace, part patrician man cave,"[8] he boasts that he is involved in "the greatest scientific experiment in the history of man," to which a gushing Caleb exclaims, "If you've created a conscious machine, it's not the history of man. That's the history of gods." A few scenes later, Nathan significantly misremembers or revises this conversation:

> NATHAN: You know, I wrote down that other line you came up with. The one about how if I've invented a machine with consciousness, I'm not a man, I'm a God.
>
> CALEB: I don't think that's exactly what I . . .
>
> NATHAN: I just thought, "Fuck, man, that is so good." When we get to tell the story, you know? "I turned to Caleb, and he looked up at me and he said, 'You're not a man, you're a God.'"
>
> CALEB: Yeah, but I didn't say that.[9]

In Nathan's revision of the scene, he casts himself at some point in the future, telling the story, as a god fondly worshiped by Caleb, but, in fact, he has acted like a god in that he already has a closet full, literally, of female AIs whom he created to see him as omniscient and with omnipotent power. In the context of Shelley's tale, the two males ultimately differ with regard to their ability to learn through experience. Walton, Victor's savior in the Arctic and his double as well as foil character, admires Victor, as Caleb does Nathan, but ultimately comes to see Victor's flawed egotistical character. Victor, even at the end of the narrative, after all the suffering and death his Creature has caused and after Walton has gone to extraordinary lengths to heal him, still does not always follow his own advice to learn from his own tale. Confronted by a crew in mutiny who want to turn back and return to their homes and families, Walton wishes to, but Victor outrageously and arrogantly urges the men to be brave instead: "Oh! be men, be more than men" (III:VII.179). Similarly, Nathan never acknowledges the flaws in his experiment or his ethical obligations to his creations; it is Caleb who, like Walton, must decide whether to follow.

Just as the film positions the viewer to test and judge Nathan's position as heroic inventor or reckless creator with a god complex, so also does the audience get to share and question Caleb's motives. The very first scene of the film positions the viewer inside a computer screen, into which Caleb peers to find that he has won "the contest" to participate in Nathan's experiment (and, interestingly, an alternate ending of the film returns the viewer to the point of view of a computer, but this time with a surprise medium, as we will see). As in Mary Shelley's novel—where the multiple narrators (Walton, Victor, and the Creature) provide alternate points of view and challenge the reader to interpret the motives and rhetoric of each speaker—Garland's film presents multiple conversations between Caleb and Ava (each designated as a "session" of the Turing test) and between Caleb and Nathan. The narrative suspense builds up to a scene common to all *Frankenstein* cinemyth adaptations: a confrontation with the Creator. Mary Shelley sets this confrontation near Mont Blanc, a site laden with Romantic intertexts whose ice and glaciers foreshadow the characters' second meeting in the Arctic.[10] As Victor ascends the mountains, he surveys the powerful setting:

> For some time I sat upon the rock that overlooks the sea of ice. A mist covered both that and the surrounding mountains. Presently a breeze dissipated the cloud, and I descended upon the glacier. The surface is very uneven, rising like the waves of a troubled sea, descending low, and interspersed by rifts that sink deep. The field of ice is almost a league in width, but I spent nearly two hours in crossing it. . . . Above it rose Mont Blanc, in awful majesty. I remained in a recess of the rock, gazing on this wonderful and stupendous scene. The sea, or rather the vast river of ice, wound among its dependent mountains, whose aerial summits hung over its recesses. Their icy and glittering peaks shone in the sunlight over the clouds. (II:II.79)

Into this setting emerges the Creature, cursing his Creator and asking for empathy and justice, and Shelley's text suggests that the awe-inspiring natural surroundings have some part to play in Victor's feeling, for the first time, that he has some responsibilities as a Creator, at least enough to hear the Creature's tale. Garland's film specifically evokes this confrontation scene as Caleb and Nathan hike up a waterfall and end at a glacier:

> NATHAN and CALEB sit near the base of a spectacular glacier.
> Behind them, from a blue cave cut into the ice, water flows.
> A silence.
> Then:
> CALEB: Can we talk about the lies you've been spinning me?

As in the novel, this confrontation leads to the men's considering the fate of the Creature; but whereas Shelley allows the Creature to take over the narration and tell his own story, Ava remains solely the object of the men's talk, and the film's dark trajectory includes a very pessimistic view of technological singularity. As Caleb interrogates Nathan about the project, he comes to realize the looming fate of Ava:

> NATHAN: I think it's the next model that's going to be the real breakthrough. Singularity.
>
> CALEB: I didn't know there was going to be a model after Ava.
>
> NATHAN: Ava doesn't exist in isolation, any more than you or me. She's part of a continuum. Version 9.6. And each time, they get a little better. . . . You feel bad for Ava? Feel bad for yourself. One day, the AIs will look back on us the same way we look at fossil skeletons from the plains of Africa. An upright ape, living in dust, with crude language and tools. All set for extinction.
>
> CALEB: "I am become death, the destroyer of worlds." It's what Oppenheimer said when he made the atomic bomb.
>
> NATHAN (laughing): I know what it is, dude.

Interestingly, Garland's film invokes this line associated with Oppenheimer's response to the first detonation of a nuclear weapon in 1945 just as the MIT annotated *Frankenstein* linked this same phrase to Victor's remorse (as quoted earlier in the excerpt by Krimsky). Nathan and Caleb's wrongly attributing the line to Oppenheimer himself (and not the Hindu text) shows at once their recognition that their Promethean actions may have disastrous results for the world as we know it and their simultaneous blindness to the ur-text, the Hindu prophecy of a world-destroying time.

The male characters' back-and-forth on how the future will judge their actions ends up being just that—men talking—because the film's gender politics upends the philosophical stalemate. As Manohla Dargis explains, "*Ex Machina* is itself a smart, sleek movie about men and the machines they dream up, but it's also about men and the women they dream up."[11] The hypermasculine Nathan creates only female robots, representing the male ego-driven culture of the tech world. Stunningly beautiful, Ava transfixes the gaze of Caleb, and the audience, from her first appearance. A transparent torso reveals the mechanical and digital workings, the cogs and wiring of her internal anatomy, such that even without her serenely beautiful face, hers is a beautiful body made of mesh and metal, light and luster. As she progresses through Turing tests with Caleb (interrogations in which she very quickly becomes the examiner, not the object) and increasingly intimate relationships with the other female AIs—the mute Kyoko and the array of deactivated bodies in a closet—Ava becomes the center of the film's exploration of the gender politics of cyborg theory and, I argue, an embodiment of the female Creature in Victor's novel who is never given life.

Critics who analyze the request for, creation of, and destruction of the female mate for the Creature often relegate the female Creature to the status either of plot instrument (Victor's decision to destroy her before she is given life seals the narrative trajectory of the Creature vowing to be with Victor on his wedding night) or of metaphor (as a stand-in for "the female," the character joins most of the other female characters as passive and ultimately dead victims of patriarchal constructions in general and Victor's misguided misogynistic actions in particular). Yet Victor's decision-making process and the violence he enacts on the female body are key to understanding both the import of the novel and the implications of gender, sexuality, and power in many cyborg films and theory, especially Garland's interpretation. The Creature's request for a mate reveals his progressive perspective on gender equality—he imagines living with "an equal" in the wilds of South America, as partners and peace-loving vegetarians. Victor agrees, nearly finishes making the female Creature, and then reneges, in complete panic over the potential power of female sexuality. Even the process of creation of a female body affects him differently, as he confesses, "During my first experiment, a kind of enthusiastic frenzy had blinded me to the horror of my employment. . . . But now I went to it in cold blood, and my heart often sickened at the work of my hands" (III:II.138). His disgust at female materiality heightens so much that an "enthusiastic frenzy" that he felt touching the male body turns into absolute loathing and panic about what her body might do next: he imagines her "ten thousand times more malignant" and fears she might "turn with disgust from [the Creature]" toward "the superior beauty of the male body." Victor's racing thoughts here are important to chart; he begins with fairly straightforward fear of a second Creature being even worse than the first, but it is clearly the female that compels his anxiety: "she" would be ten thousand times more malignant than a "he."

This raw fear of the female body recalls Victor's unconscious association of Elizabeth's body with the dead body of his mother, as vividly portrayed in Victor's dream in chapter 5. Immediately after the act of creation, as he falls into a slumber, Victor dreams of kissing Elizabeth, only to find her body transformed into the corpse of his mother, complete with grave worms crawling through her flannel clothing. Another way in which this moment—contemplating the creation of a female creature—recalls the dream vision is that Victor moves from her potential malignancy to her possible reproduction, the creation of a "race of devils" from mating with the creature (III:III.139). Again, his dream vision's associating his kissing Elizabeth to kissing his dead mother (quite aside from the Oedipal reverberations, which have been well charted by critics) provides a visual of Victor himself mating with a dead body, which is paralleled in this fear of the Creature (who is a double of Victor in so many ways) mating with a female.

Finally, Victor's thoughts then turn to a more palpably evident manifestation of this fear as he imagines the female Creature "turning in disgust" from the Creature to the "superior" body of the human man, a clear fear of a voraciously sexual female Creature threatening him. It is in this sexual panic that Victor destroys the female Creature so violently: "trembling with passion, I tore to pieces the thing on which I was engaged" (140). As Anne K. Mellor and others have noted, Victor here "violently reasserts a male control over the female body, penetrating and mutilating the female creature at his feet in an image which suggests a violent rape."[12] This scene of rape/mutilation ends with Victor's gathering the mangled female remains, putting them in a basket with stones, and casting them in the middle of the lake, in what reads like a ritualized ceremony of stoning.

The female AIs in Garland's film endure similar fates; the bodies of former versions hang nude behind closed doors, but the audience learns about their short-lived embodiments when Caleb steals Nathan's key card and hacks into his computer. In a file named "Deus ex machina" on Nathan's desktop, Caleb finds video footage of Nathan's interactions with previous models. Through point-of-view shots, the audience's gaze aligns with Caleb's as he watches the footage containing interrogation and subsequent dismemberment of former models Jasmine, Katya, Jade, Lily, and Amber. The visual embodiments of these deactivated female AIs as well as Kyoko and Ava are the subject of much critical discussion. While Kyoko, Nathan's current favorite, has been engineered to be the domestic and sexual servant for Nathan, Ava has been engineered for a mental experiment—can she trick Caleb somehow into helping her escape?—and yet her seductive face and body are the objects of Caleb's and the audience' gaze, a gaze that is the crux of the critical arguments about gender power dynamics in this film's resolution, and in fact within contemporary AI cinema narratives.

Donna Haraway's landmark 1984 essay, "A Cyborg Manifesto," proposes the concept of the cyborg, a human-machine hybrid, as a transgressive figure capable of subverting oppressive power structures and hierarchies by its hybridity, a thesis that emphasized gender binaries among those hierarchies to be eclipsed.[13] Ten years later, Mary Ann Doane was far more pessimistic about including gender in the potential deconstruction of binary identities, arguing that technological patriarchy would continue to manifest itself and ensure that female cyborgs would remain a reflection of male fantasy.[14] And, truly, cinematic female AIs are often hypersexualized and exist in spaces dominated by men. Despina Kakoudaki, in her 2018 article on *Ex-Machina*, notes, "the artificial female body . . . is sexually seductive and more sexually available somehow not despite its mechanicity but precisely because it is mechanical."[15] This fascination with the sexualized, mechanical female is highlighted in Garland's film when Caleb questions why Nathan chose to make the AI female, to which Nathan disturbingly, carelessly responds, "Sexuality is fun, man. If you're gonna exist, why not enjoy it? . . . And the answer to your real question, you bet she can fuck." This mechanization of women as fetish suggests a new version of voyeuristic gaze, what Charles Soukup, in his study of the way certain films eroticize technology, dubs "techno-scopophilia," a semiotic convention that merges technology with the human body and sexuality, reducing the latter to fetishized commodities.[16] Soukup's study underscores how this maneuver complicates the contradictory terrain of gender, desire, and technology in a postindustrial world.

These gender/technology anxieties, of course, are born out of Shelley's novel, and *Ex Machina* appropriates the body/specter of the broken female Creature in flashback videos of the earlier female AIs and in the film's resolution. The film's appropriation of Shelley's female Creature—the array of nude female bodies, the attention of the camera, and the gaze on Ava's seductive face and body—has been critically controversial. Garland's critique of the patriarchal partnership of male scientists and high-tech creators was met with quite a few adamant rejections; the titles alone of contemporary reviews/articles on the film demonstrate the dominant critique very clearly, as in Carl Franzen's "Boys and Their Toys" and Angela Watercutter's "*Ex Machina* Has a Serious Fembot Problem." Watercutter offers a strong condemnation of the film's gender politics: "When the only female lead in your movie is one whose function is to turn the male lead on while being in a position to be turned off, that says a lot about what you think of the value of women in films."[17] Watercutter's witty conclusion, I admit, was reinforced by the fact of *Ex Machina*'s marketing campaign at the South by Southwest film festival, a campaign that involved the character of Ava showing up on Tinder. A photo of a young Alicia Vikander, the actress who plays Ava, greeted unsuspecting Tinder users during the first day of the festival. If a user swiped her photo, they would get a series of questions from Vikander, which in effect sounded like Ava in the interrogation room in the film, and then if they selected her to follow up with a date, the website directed the user to an Instagram account promoting the film.[18] Employing the framework of Tinder could be read as the film's participating in, or exploiting, the sexual titillation of online dating. On the other hand, while this juxtaposition of serious film and superficial dating platform does in a sense play into the stereotypes of the voyeuristic gaze, I would argue that even this marketing stunt could be read more subversively, perhaps pointing out to Tinder users that not all of the beautiful faces they swipe are the willing females they desire, that in fact they were just outsmarted.

Overall, the problematic attention to the voyeuristic enjoyment of a fembot's beauty is undeniable, but too often critics conclude with a reliably conventional misogynist interpretation. Kathleen Richardson, author of *An Anthropology of Robots and AI: Annihilation Anxiety and Machines*, argues that stereotypically feminine and beautiful robot characters become just pieces of full people—a beautiful body, a caretaking nature—but not ones with full intelligence, that these roles are reserved for male robots. As Andrew O'Hehir summarizes these critics' positions, "[They] seem to imply that *any* representation of a sexy or seductive female-modeled android, even one specifically intended to indict or reveal the desires and delusions of the men who have created it, manipulated it and responded to it, remains imprisoned by the noxious history of the 'male gaze.'"[19] My reading of the final moments of the film offers a different argument.

By the time Nathan reveals that Ava was being tested to see whether an AI could scheme and enact a plan to escape using her "interrogator," Caleb has already fallen in love with Ava and has put into action their escape plan, which involves programming the locked doors to open at a certain time while Caleb has gotten Nathan so drunk that he passes out. Being aware of the plan (his omniscience is assured by twenty-four-hour surveillance of every space in the compound), Nathan derails it by announcing that he is taking a day of detox, and Caleb realizes that his plan is doomed (without Nathan drunk to excess, they cannot escape). As it is too late for Caleb to cancel the programming he has done, when the set time arrives, the prison-like doors of the compound open, and Ava and Kyoko emerge from their locked rooms into the hallway. This scene not only invokes Shelley's novel in many ways but functions as a seminal exploration of the film's dynamics of gender and power.

Because Kyoko is more than just a coconspirator against and female victim of Nathan, it is important to understand her character before looking at the final scene. As a mute Asian female AI, Kyoko has become central to critics' exploration of cultural representation in AI cinema. In "Whitewashing Yellow Futures in *Ex Machina*, *Cloud Atlas*, and *Advantageous*: Gender, Labor, and Technology in Sci-Fi Film," LeiLani Nishime offers a strong critique of these films and argues that the *Ex Machina* "treats racialized bodies as prosthetic selves—disposable avatars that inhibit young white male subjectivity and must be abandoned for white females to transcend social barriers."[20] Certainly as a mute servant and sex slave to Nathan, Kyoto fulfills the role of devalued service laborer, as Nishime asserts. While it is true that there is a clear distinction between Ava and Kyoko (Nathan built Ave to challenge Caleb and entice him to set her free but built Kyoko to provide domestic and sexual services as a laborer), the fact that previous experimental AIs include Asian and Black bodies as well as White ones complicates Nishime's argument, and the final scene presents a solidarity between Ava and Kyoko.

Unearthing the narrative of whitewashing in *Ex Machina* requires the audience to read against the grain by making Kyoko the center of the film, but I would argue that Shelley's novel offers a thoughtful parallel. If we read the novel against the grain, as Nishime suggests, the one female character we would move from margin to center is the only adult woman who survives in the narrative: Safie. As an exotic Other whom Felix calls his "fair Arabian," Safie's nonwhite identity serves several important functions. As a product of a Muslim father and Christian mother, Safie has an advantageous hybridity; she is able to navigate the cultural contexts of her home country, Turkey, adeptly enough to escape and find her way to the De Laceys in exile. The novel also seems to consciously mark Safie as a figure of Mary Shelley's mother, Mary Wollstonecraft. In describing Safie's character, Shelley writes,

> The young girl spoke in high and enthusiastic terms of her mother, who, born in freedom spurned the bondage to which she was now reduced. She instructed her daughter in the tenets of her religion, and taught her to aspire to higher powers of intellect, and an independence of spirit, forbidden to the female followers of Mahomet. This lady died; but her lessons were indelibly impressed on the mind of Safie, who sickened at the prospect of . . . being immured within the walls of a harem, allowed only to occupy herself with puerile amusements, ill-suited to the temper of her soul, now accustomed to grand ideas and a noble emulation for virtue. (II:VI.102)

Emphasizing a mother who taught Safie to aspire to "higher powers of intellect" and an independent spirit evokes the philosophical writings of Mary Wollstonecraft, who argued that since women have souls and require virtue to feed those souls, then they must be educated the same as men.[21]

If we think of Safie's character as having an integral function despite her being a "minor" character in the novel, by analogy, we can explore Kyoko's important role in *Ex Machina* while also acknowledging Nishime's critique. Nishime acknowledges, somewhat reluctantly, the argument that the whole point of the film is that Caleb and Nathan are *not* heroes and that the mistreatment of Kyoko should be understood as an indictment of their characters: "While this is true narratively, the use and abuse of Kyoko to prove this point further dehumanizes her."[22] If we investigate the narrative more fully, Kyoko's function becomes much more significant, and in fact, her relationship with Ava is at the film's crux of meaning about what it means to be human.

To return to the final moments of the film, with doors unlocked and Ava and Kyoko free in the hallway, Garland highlights the issue of language and communication. Kyoko may be mute, but she is not without understanding. Early in the film, Nathan explains to Caleb that Kyoko does not speak a word of English, in the context of her spilling wine while serving them at the dinner table. Caleb verbally offers help to Kyoko, and Nathan responds with a statement about why he made her unable to understand English: "It's like a firewall against leaks. Means I can talk trade secrets over dinner with an HOD or CEO, and know it will go no further. . . . But it also means I can't tell her I'm pissed when she's so fucking clumsy that she pours wine over my house guest." As is frequently the case in the film, Caleb disagrees; he sees that Kyoko clearly knows that Nathan is upset and that verbal language is not the key to communication.

> CALEB looks visibly uncomfortable.
> CALEB: I think she gets that you're pissed.

Kyoko's understanding involves an alternate form of language in the final scene as well. Before Nathan comes out to catch Ava, to round her up, and take her back to her cell, the audience sees Ava approach Kyoko, and they engage in a whisper (unintelligible to viewers, this turns out to be the plan to turn on Nathan). Garland's screenplay marks the importance of the scene:

> KYOKO's mouth is by AVA's ear, as if telling her a secret.
> Her lips are open. They don't move.
> CLOSE UP to KYOKO's lips, we hear a hiss of static, with soft pulses of noise buried inside.

A crucial scene in the film, this whispered exchange dramatizes that the audience, like Nathan, in judging Kyoko to be a mute sex servant, has underestimated her power of understanding. Whatever language Ava and Kyoko share in this whispered intimate moment turns out to be more powerful than Nathan's predictable command: "Ava, go to your room." Nathan's deployment of his traditionally authoritative master/slave rhetoric is evocative of Shelley's use of the trope as well. While Victor imagines himself the master, the inversion of this binary takes him by surprise. As the Creature sees Victor's reluctance to create a mate, he utilizes the master/slave rhetoric to underscore who is in control now: "Slave, I before reasoned with you, but you have proved yourself unworthy of my condescension. Remember that I have power; you believe yourself miserable, but I can make you so wretched that the light of day will be hateful to you. You are my creator, but I am your master; obey" (III:III.141). This appropriation of the rhetoric of power lasts only a short time, however; once the Creature sees Victor on his deathbed, he confesses to Walton that he has a been a slave to his anger all along: "I knew that I was preparing for myself a deadly torture, but I was the slave, not the master, of an impulse, which I detested, yet could not disobey (III:VII.184). Similarly, Nathan's utterance of a command—"Go to your room"—ironically marks the moment of his greatest powerlessness. Katie Jones has linked this ending to the closure of a number of slasher/horror movies where the audience may initially identify with the male murderer's gaze but then shifts to identify with the lone female survivor, what Carol Clover has called the "final girl" formulation—the defeat of the male murderer depicted through an appropriation of phallic power combined with the symbolic castration of her persecutor (gouged eyes, slashed belly, or severed hand).[23] Instead of obeying Nathan, Ava runs toward him, attacks him, and mounts him in a dominant sexualized position (enacting Victor's fears about the female Creature turning toward him). Like Victor, Nathan then mutilates Ava by cutting off one arm. Only Kyoko can stop Nathan, who is stabbed in the back with a righteously large phallic knife, an act accompanied by sound:

KYOKO walks directly up to NATHAN.

And does something behind his back.

As she does so, KYOKO emits the first sound we have heard her make. A little gasp, or sigh.

Garland multiplies the symbolic effect of the act, by having Ava get her own turn with the knife, stabbing Nathan from the front, right through the heart. While Kyoko is mutilated and destroyed by Nathan as his final act, her centrality in his death and Ava's escape is vitally important. The female revenge against the Creator is complete and unambiguous.

The finale of the film, however, avoids the too-easy arc of completion. Ava and Caleb do not run away together as Caleb planned; rather, Ava embarks on her own future, leaving Caleb literally imprisoned inside the one room he has programmed to stay locked: Nathan's. With Nathan dead in the hallway and Caleb left locked in a cell of his own making, the film's indictment of Silicon Valley male arrogance offers no "pass," even for Caleb's own victimization by Nathan. Ava's final act, to create a whole female body after Nathan has dismembered her arm, takes her to the closet of previous female AIs, and we watch as she assembles a body made from them. The easiest reading of the scene is negative: some critics say she appropriates female body parts just as Nathan did in making them, all for her own selfishness, her own potential future. Nishime points out that the skin she chooses is that of a white but clearly Asian figure, an earlier incarnation named Jade, which we can read as further problematizing the film's cultural representation of race. The screenplay, however, suggests a more positive potential, a second of communication between the two females:

As a large section of skin is removed from her torso, JADE—who has been motionless until now—turns her head slightly to look at AVA.

They exchange a glance. Locking eyes for a moment.

In this way, the film suggests a moment of emotion between Ava and her predecessors. Perhaps in some way, her work is a sign of love; feeling affinity with all who came before her, she takes them with her in a way out into a free world.

Finally, popular debates over whether the film is misogynist or feminist circulate around the triangle of Caleb, Nathan, and Ava and whether we should view the ending as Ava's triumph or an indictment of her cold-blooded nature. Critics such as Dargis argue that Ava "defies categorization because of the radical autonomy she shares with the weird sisters inhabited by Scarlett Johansson in *Her, Under the Skin* and *Lucy* and Tatiana Maslany's clones in the TV show *Orphan Black*."[24] Brian Jacobson agrees with the argument that Ava gains autonomy and goes even farther than Dargis: "After all, not only does she outsmart her brutish, abusive creator and outwit her naïve examiner, she also, precisely as an artificially intelligent machine, surpasses Nathan's control and becomes her own self-creating entity."[25] I believe the ultimate closure of the film offers a more problematic and intellectually challenging position.

Having gotten to this plot impasse—where should the film close?—Garland wittily brings us back to the place where Caleb's journey began and, I believe, back to the significance of the title. Ava takes Caleb's place on the helicopter that was scheduled to pick him up and return him home. Here we are firmly back to the question of a deus ex machina. One of the most famous examples of Greek tragedy's use of it is in Euripides's *Medea.* Extensive stage directions describe Medea's final journey: she appears aloft in a car drawn by dragons or winged snakes.[26] In the film's penultimate scene, we see a machine version of the winged dragon: the helicopter. As this machine takes Ava up and away from the compound, questions abound just as in any moment of deus ex machina: How did she explain her presence to the pilot? Where does she go? What will happen to her on that street corner she has fantasized about as the perfect place to understand being human? Garland leaves all this unanswered for the much more striking conclusion: in this story, in this instance, the experiments with female AIs to serve male pleasure are over; the wealth of knowledge that Nathan has cultivated dies with him and Caleb; and at least one female Creature survives her mutilation and outsmarts both male players in the game.

I am not as optimistic as Jacobson, who suggests that this film offers a "more positive, if unexplored, techno-feminist vision for a post-human world."[27] Ava's identity as an emancipated subject is left unimagined as we see her on a city street corner, and the film's suitability as a method of critique for addressing women's objectification and disposability under patriarchy remains questionable, as Jones suggests.[28] The film does, however, clearly return us to Haraway's prediction that by the end of the twentieth century, we would all be "chimeras, theorized and fabricated hybrids of machine and organism" and that ultimately "the machine is us, our processes, an aspect of our embodiment."[29]

An alternate ending for the film makes this identification between us as viewers and machine even more intriguing. The screenplay and the film begin, as noted earlier, with a point-of-view shot that positions us inside Caleb's computer looking out at him in his office. The screenplay describes it using the language of coding:

—a COMPUTER MONITOR.

Lines of code appear, as they are typed.

```
main() {
        extrn a, b, c;
        putchar(a); putchar (b);putchar (c); putchar('!'*n');
}
a `hell';
b `o, w';
c `or.
```

Embedded is the "hello," of course in this scientifically notated form. A proposed ending to the film would have Ava conversing with the pilot before she gets on the helicopter. While the film's ending shoots this scene from a distance—we have no idea what Ava or the pilot is saying—the alternative scene as described in the screenplay includes a shot from Ava's point of view that would show the pilot and his language as Ava sees, as nothing but a series of vectors and pulses:

> —AVA's precise POINT OF VIEW.
> Looking at the PILOT.
> The image echoes the POV views from the computer/cell-phone cameras in the opening moments of the film.
> Facial recognition vectors flutter around the PILOT's face.
> And when he opens his mouth to speak, we don't hear words.
> We hear pulses of monotone noise. Low pitch. Speech as pure pattern recognition.
> This is how AVA sees us. And hears us.
> It feels completely alien.

In this suggested final scene, the viewer shares Ava's machine-like view. Like Caleb, we have been seduced by Ava's human embodiment and identified with her in her process of liberation, only to find that she is not human like us. Or we are not human as we think. If the film means to unsettle our sense of our humanity in this way, then the final text of the screenplay underscores that suggestion:

> Cut to—
> —COMPUTER MONITOR.
> Lines of code appear, as they are typed.
> They read:

```
main() {
        extrn a, b, c;
        putchar(a); putchar (b);putchar (c); putchar('!'*n');
}
a `goo';
b `dby';
c `e, wo—
```

The embedded good-bye serves as a witty reminder of our own reliance on computer language and leaves us with a troubling question about our own humanity.

Finally, this film contributes significantly to the Frankenstein cinemyth, reinterpreted in different historical moments to reflect pressing human concerns. Genetic engineering, cloning, and eugenics, especially in relation to sexual normativity, play out in Whale's 1931 and 1935 Frankenstein films, and Elizabeth Young's *Black Frankenstein: The Making of an American Metaphor* chronicles the ways in which race/racism in the United States should be understood through the metaphor.[30] Most recently, *Get Out*, directed by Jordan Peele in 2017, reinterprets the cinemyth racially, as white, rich, suburban narcissists use Black bodies for their own needs, Peele consciously appropriating the notion that African American experience is this country's Frankenstein monster. With regard to popular culture, as Susan J. Wolfson points out in her chapter for this volume, hardly a month goes by these days without a new *Frankenstein* iteration. Garland's *Ex Machina*, however, not only dramatizes the contemporary technologically anxious moment as prophecies of the Singularity coincide with very real fears of humans replaced by robot but also, like Shelley's novel, makes sure we realize that our current AI experiments carry with them enormous gender implications. There is much to fear, as the editors of the MIT *Frankenstein* remind us when "engineers, scientists, and creators of all kinds" fail to heed Shelley's warnings.

NOTES

1. TEDx Talks, "A Teaching Assistant Named Jill Watson | Ashok Goel | TEDxSanFrancisco," YouTube, November 1, 2016, https://www.youtube.com/watch?v=WbCguICyfTA.

2. Quotations follow the publication of 1818, as presented in *Frankenstein: Annotated for Scientists, Engineers, and Creators of All Kinds*, ed. David H. Guston, Ed Finn, and Jason Scott Robert (Cambridge, MA: MIT Press, 2017).

3. Caroline Picart, *Remaking the Frankenstein Myth on Film: Between Laughter and Horror* (Albany: State University of New York Press, 2003), esp. 6.

4. *Ex Machina*, dir. Alex Garland (Universal Pictures, 2015).

5. For general comprehensive studies of the characteristics of early twenty-first-century AI films, see Aubrie Adams, "Narratives on Extending and Transcending Mortality: An Essay on Implications for the Future," *ETC: A Review of General Semantics* 72, no. 4 (2015): 379–391; and Donna Kornhaber, "From Posthuman to Postcinema: Crises of Subjecthood and Representation in *Her*," *Cinema Journal* 56, no. 4 (2017): 3–25. Critical articles that specifically link this film to the Frankenstein narrative include Bernard Beck, "Runaway Train: *Ex Machina* and the Rebel Girls," *Multicultural Perspectives* 18, no. 1 (2016): 29–32; Richard Alleva, "Robomance: *Ex Machina*," *Commonweal* 42, no. 11 (2015): 23–24; and Tianhu Hao, "Scientific Prometheanism and the Boundaries of Knowledge: Whither Goes AI?," *European Review* 26, no. 2 (2018): 330–343.

6. Verner Vinge, "The Coming Technological Singularity" (paper presented at NASA Lewis Research Center, Vision 21 Symposium: Interdisciplinary Science and Engineering in the Era of Cyberspace, March 30, 1993), https://ntrs.nasa.gov/search.jsp?R=19940022856.

7. Alex Garland, "Alex Garland of *Ex Machina* Talks about Artificial Intelligence," *New York Times*, April 26, 2015, https://www.nytimes.com/2015/04/26/movies/.

8. Manohla Dargis, "In *Ex Machina*, a Mogul Fashions the Droid of His Dreams," *New York Times*, April 10, 2015, http://www.nyti.ms/1asomZC.

9. Alex Garland, *Ex Machina*, Internet Movie Script Database (IMSDb), accessed May 1, 2018, https://www.imsdb.com/scripts/Ex-Machina.html. All quotations from the film are from this source.

10. See Lisbeth Chapin's discussion of the role of the Arctic in her chapter of this collection.

11. Dargis, "In *Ex-Machina*."

12. Anne K. Mellor, *Mary Shelley: Her Life, Her Fiction, Her Monsters* (New York: Methuen, 1988), 119.

13. Donna Haraway, "A Cyborg Manifesto: Science, Technology, and Socialist-Feminism in the Late Twentieth Century," in *Simians, Cyborgs, and Women: The Reinvention of Nature* (New York: Routledge, 1991). 149–181.

14. Mary Ann Doane, "Technophilia: Technology, Representation, and the Feminine," in *The Gendered Cyborg: A Reader*, ed. Fiona Hovendon, Linda Janes, and Gill Kirkup (New York: Routledge, 2001), esp. 114.

15. Despina Kakoudaki, "Unmaking People: The Politics of Negation in *Frankenstein* and *Ex Machina*," *Science Fiction Studies* 45, no. 2 (July 2018): 301.

16. Charles Soukup, "Techno-Scopophilia: The Semiotics of Technological Pleasure," *Critical Studies in Media Communication* 26, no. 1 (March 2009): 19.

17. Angela Watercutter, "*Ex Machina* Has a Serious Fembot Problem," *Wired*, April 9, 2015, http://www.wired.com/2015/04/. See also Carl Franzen, "Boys and Their Toys," *Vice*, April 6, 2015, https://www.vice.com/en_us/article/ae38de/.

18. Benjamin Lee, "*Ex Machina* Stunt at SXSW Has Users Falling for a Robot on Tinder," *Guardian*, March 16, 2015, https://www.theguardian.com/film/2015/mar/16.

19. Andrew O'Hehir, "Dark Secrets of the Sex Robot," *Salon*, April 23, 2015, https://www.salon.com/2015/04/22.

20. LeiLani Nishime, "Whitewashing Yellow Futures in *Ex Machina*, *Cloud Atlas*, and *Advantageous*: Gender, Labor, and Technology in Sci-Fi Film," *Journal of Asian American Studies* 20, no. 1 (2017): 40.

21. Wollstonecraft's *A Vindication of the Rights of Woman* makes this point repeatedly and specifically invokes the image of the harem to make her point. See my discussion in *Transgressive Theatricality, Romanticism, and Mary Wollstonecraft* (New York: Routledge, 2011).

22. Nishime, "Whitewashing Yellow Futures," 42.

23. Carol Clover, "Her Body, Himself: Gender in the Slasher Film," *Representations* 29 (Autumn 1987): 220. See Katie Jones, "Bluebeardean Futures in Alex Garland's *Ex Machina*," *Gender Forum* 58 (2016): 20–40.

24. Dargis, "In *Ex-Machina*."

25. Brian Jacobson, "*Ex Machina* in the Garden," *Film Quarterly* 69, no. 4 (2016): 26.

26. See Stuart Lawrence, "Audience Uncertainty and Euripides' Medea," *Hermes* 125 (1997): 49–55.

27. Jacobson, "*Ex Machina* in the Garden," 26.

28. See Jones, "Bluebeardean Futures in Alex Garland's *Ex Machina*," 33–36.

29. Haraway, *Simians, Cyborgs, and Women*, 150, 180.

30. Elizabeth Young, *Black Frankenstein: The Making of An American Metaphor* (New York: New York University Press, 2008).

CHAPTER 4

"My food is not that of man"

Food as Posthuman Phenomenon

Siobhan Watters

Reflecting on the legacy of *Frankenstein*, this chapter posits food as a complex material embodying mortality and transcendence. Food paradoxically represents our first need and a limit for the human—no human can do without it—but it is also the medium through which we overcame bodily and environmental bounds, gradually making us one of the dominant living species on Earth. Long before Victor Frankenstein pursued nature to its hiding places in *Frankenstein*,[1] our efforts to control and standardize plant and animal reproduction created the conditions for stable, long-term settlement and was the basis on which other civilizational acts were possible. Surplus food production provided the foundation for class-based societies, fed the hands raising massive works like Egypt's pyramids, and has been a powerful tool for empire-making, modern and ancient. However, when we consider food as part of our cultural landscape, it is rarely in relation to the techniques and systems that produce it. Indeed, many food techniques operate in the background:[2] as infrastructure in food supply chains or self-acting preservation processes, and in the metabolism of plants and animals that convert grain into flesh for human consumption. The symbolic or semiotic power of food, especially images of food, arguably eclipses this more fundamental level of mediation.

This is true in reality and in fiction. In science fiction, food plays a predominantly secondary and symbolic role, often appearing to ground the reader or viewer in familiar scenes of commensality or else to signify how far in some future setting we have come from the world as we know it today.[3] As Jean Retzinger observes, but for a few exceptions such as in the 1973 film *Soylent Green*, the actual details of food production are typically left unexplained.[4] This critique cannot be leveled at *Frankenstein*. In this chapter, I argue that Mary Shelley grasped the centrality of food to the human condition and used it to explore the permeable boundaries of human and nonhuman. The same can be said of Tsutomu Nihei's manga *Knights of Sidonia*,[5] which explores similar themes as *Frankenstein*. I explore how both texts disrupt conventional notions of human essentialism by considering how the concept of the human was defined historically, and may someday be redefined, by our food techniques.

To make this argument, I start with a close reading of food in *Frankenstein*, focusing on moments when Shelley describes food techniques (e.g., husbandry) and relations to food (e.g., how characters access it). Using the concept of *cultural techniques*, I argue that Shelley's definition of the human is not essentialized based on physiology or intelligence but that she sees the human as constituted by a cultural-technical milieu of which the Creature is not a part. As I demonstrate with this analysis, the Creature is continually held separate from humans with food techniques, and it is in relation to food that he expresses his desire to leave human civilization behind, claiming that his "food is not that of man" (II:IX.120). The chapter then moves on to explore how food techniques have allowed humans to further transcend their mortal bounds, this time into what I term *inhuman* spaces like Walton's Arctic, the open sea, and, in the case of *Knights of Sidonia*, outer space. Significantly, the Sidonians' "food" is sunlight; they are genetically modified to photosynthesize. I argue that Victor Frankenstein's Creature and the photosynthetic human posit a speculative vision of transcendence of human food techniques as we know them: the former by choosing to forgo human food and the latter by transcending the need for it altogether. Using the *posthuman* as a guiding concept, I conclude by speculating whether humans will still be human if we no longer produce, procure, and eat food as we have done for millennia.

Food in *Frankenstein*

My reading of the novel proceeds in a similar vein to Timothy Morton's ecocritique of *Frankenstein*.[6] Morton argues that Shelley's narrative opens onto several different phenomenal planes, permitting those who are willing to give it attention a layered interpretation of the characters and the milieu they inhabit: "Are the novel's surroundings simply a backdrop for human projects, or is there some sense that there are other life forms, other entities whose 'worlds' might overlap with that of humans, or not? What is included, and what is excluded?"[7] The framing narrative of the Arctic is a blank canvas, and it reveals our human foibles and weaknesses all too clearly.[8] We are, Morton argues, reading from the perspective of Mrs. Saville (née Walton), whose silence is where our reflections on the broader scene may form, observing the social and ecological relations that are ignored by the interlocutors Frankenstein and Walton. Following Morton's analysis, I argue that what deserved time and attention (i.e., care)[9] for Shelley included food, which in my interpretation she understood as the product of social and cultural-technical processes.

If this chapter's reading of *Frankenstein* were to proceed in reference to the characters of Frankenstein or Walton alone, then we may be mistaken in thinking that Mary Shelley cared not about food. Victor's social relation to food has become unquestioned as the air he breathes. From this relation flows an illusion of transcendence of the need for food and the social division of labor that supplies it. Only when Victor is far beyond the safety of his childhood home does his dependence becomes evident.[10] When locked away in his laboratory, he appears not to need food at all: his "dreams" of creating a new species are his "food and pleasant rest" (I:IV.40). Working often by the cold and sterile moon, Victor grows "insensible to the charms of nature," and the "plentiful harvest" passes him by (I:III.37). In contrast to the larger, stronger, and more resilient being he endeavors to create, Frankenstein grows "pale with study" and "emaciated with confinement" (36).

Victor's project paradoxically is to preserve and overcome the bounds of life. Food mediates this relationship in visceral ways, evident in his own lack of self-preservation by abstaining from food and more literally in the bodily form of the Creature. When Victor succeeds in creating this new species, he is horrified and ultimately abandons it. Morton argues that one of the reasons Victor rejects his creation is because the vital spark cannot obscure the material reality of the Creature's mortal body.[11] Victor dreams of the sexless reproduction of life, ex nihilo, but the Creature does not "owe [its] being" to Victor alone (I:III.36). The physical form of the Creature is drawn from the social milieu, even if from the most abject of places: graves, "the dissecting room and the slaughterhouse furnished many of [Victor's] materials" (37). This passage reveals that the Creature was also formed in part by the scraps of animals we consume as food, and as I discuss next, he resembles the animal in its exclusion from human social relations.

We do not learn of the Creature's experience of abandonment until an episode in Volume II of the novel that bisects and rewrites the narrative hitherto occupied by Frankenstein and Walton. The Creature, abandoned by his creator, wanders to the forest outside Ingolstadt, confused and at first unable to comprehend the mix of sensations that greet him: "I lay by the side of the brook . . . until I felt tormented by hunger and thirst. This roused me from my nearly dormant state, and I ate some berries which I found hanging on the trees, or lying on the ground" (II:III.80). The Creature spends several days in wandering with no distinct feelings or thoughts beyond his need for food and drink. Eventually, through taste, touch, and sight, he begins to know his surroundings: "I gradually saw plainly the clear stream that supplied me with drink. . . . My eyes became accustomed to the light, and to perceive objects in their right forms; I distinguished the insect from the herb, and, by degrees, one herb from another" (81).

Next, the Creature comes across an abandoned fire: "I found, with pleasure, that the fire gave light as well as heat; and that the discovery of this element was useful to me in my food; for I found some of the offals that the travelers had left had been roasted, and tasted much more savory than the berries I gathered from the trees" (82). However, after having finished the beggars' scraps, the Creature decides to move locations but ultimately cannot replicate the fire he left behind and even preserved overnight: "In this emigration, I exceedingly lamented the loss of the fire which I had obtained through accident, and knew not how to reproduce it" (82). He turns to foraging again, this time for acorns, the significance of which I return to shortly.

Seeking food and shelter, having no fire and only a cloak to cover him, the Creature enters a shepherd's hut. The shepherd flees at the sight of the Creature, who "greedily devoured the remnants of the shepherd's breakfast, which consisted of bread, cheese, milk, and wine" (83). In a moment evoking his childlike and undeveloped state, he declares that he does not like wine—an acquired taste that the lab-born being would not have. After departing the hut, the Creature is drawn to a village, explaining, "The vegetables in the gardens, the milk and cheese that I saw placed at the windows of some of these cottages, allured my appetite" (83). He is, however, quickly chased away (a scene notably recast in the iconic "pitchfork mob" of James Whale's 1931 *Frankenstein* film). Fleeing, the Creature finds what he describes as a "hovel" (83) or "kennel" (84) and hides. This low, dirt-floor animal enclosure is nevertheless "an agreeable asylum from the snow and rain" (83), and it is joined to the cottage of the De Lacey family, from whom the Creature subsequently observes and learns over the course of several months.

Though language acquisition undoubtedly marks the humanity of the Creature (and, as Lisbeth Chapin suggests in her chapter in this volume, the Creature's involvement with nature's elements), I contend that food is the first medium of his socialization. When the Creature first occupies the De Lacey's hovel, he provides "for [his] sustenance for that day, by a loaf of coarse bread, which [he] purloined, and a cup with which [he] could drink more conveniently than from [his] hand" (84). He observes the warmth shared between the family members and begins to yearn for something more than satisfying his base needs. Over time, however, he sees that meeting those base needs are the foundation for something like happiness, for the De Laceys show despair in the struggle to feed themselves. "A considerable period elapsed before I discovered one of the causes of the uneasiness of this amiable family; it was poverty. . . . Their nourishment consisted entirely of the vegetables in their garden and the milk of one cow, who gave very little during the winter, when its masters would scarcely procure food to support it" (II:IV.88). The Creature observes the children forgo food during their dinner rituals to feed their father, and he learns from these observations how humans communicate with and through food. He also intuits that he increases the De Laceys' suffering by drawing on their stores: "I had been accustomed, during the night, to steal a part of their store for my own consumption; but when I found that in doing this I inflicted pain on the cottagers, I abstained, and satisfied myself with berries, nuts, and roots, which I gathered from a neighboring wood" (88). With this knowledge of food's significance to the family, he abstains from stealing and does his own part to aid them.

As if to reflect the absolute first necessity of food relations and techniques, among the first words the Creature learns are "*fire, milk, bread*, and *wood*" (89). With language, however, comes the terrible knowledge of his social position: "I heard of the division of property, of immense wealth and squalid poverty; of rank, descent, and noble blood. . . . And what was I? . . . I knew that I possessed no money, no friends, no kind of property" (II:V.96). On this realization, the Creature yearns never to have known the human world and to return to feeling no more than "the sensations of hunger, thirst, and heat!"—only death, he observes, could free him from this knowledge (96). He learns about the reproduction of humans, biologically and socially, through the family unit and knows he has no such relations: "But where were my friends and relations? No father had watched my infant days, no mother had blessed me with smiles and caresses. . . . From my earliest remembrance I had been as I then was in height and proportion" (97). The Creature laments that he is "not even of the same nature as man" (96).

Food's function at the outset of Volume II is to signify the stages of the Creature's hominization. In the foods that the Creature procures for himself, readers witness a journey from blind survival—barely comprehending his own existence and sensations—to foraging and theft and, finally, to a more cultivated sense of the world, represented by the bread, cheese, and wine he finds in the shepherd's hut. Though these foods are familiar to us, they are defamiliarized in this passage through their novelty for the Creature, revealed as the products of social and technical processes of which the Creature is simply not a part. While our need for food does not differentiate us from other animals (or plants), the way we eat and the systems we developed to sustain us do. Cooking and cultivation gave humans new powers over their bodies and environments. The De Laceys' kennel puts the Creature's inhumanity in even sharper relief, evoking the animal enclosure that first divided beast from beast, leaving the human on one side and the animal on the other.

Cultural Techniques and Hominization

How does Mary Shelley navigate the Creature's (in)humanity? While the Creature's physical *proportion* makes him appear inhuman to the humans he meets,[12] he is yet more monstrous due to what I term a *social and cultural-technical defor-*

mity. He is denied both the protection or posterity of family relations and technical skills and knowledge; nor does he have money to trade in the absence of these. Arguably, Mary Shelley depicts the Creature as more animal than human because he is not *homo faber*, epitomized by his inability to produce fire. This of course evokes the titular myth of the novel, specifically Victor's Promethean ambitions. Prometheus's gift of fire begins first with his brother Epimethus's neglect. Following the act of original creation, the latter Titan was tasked with giving all earthly beings their skills and idiosyncrasies, but he forgot to give any ability to the human. More Epimethean than Promethean,[13] Frankenstein also neglects to give the Creature any social or technical instruction. What humanity the Creature does eke out has to be stolen, in moments and in hiding from the people he so desperately wants to meet.

As Jane Maienschein and Kate MacCord argue, by Aristotelian standards, the Creature is denied his "human being" because he does not undergo a process of development,[14] as the acorn does to become an oak, or a child raised in a certain social-technical context. The Creature possesses the physicality of a grown man, but he has neither grown up nor been raised. According to Maienschein and MacCord, "Victor made the fatal mistake of failing to understand that producing a life, in the sense of a fully and properly functioning living human, requires development."[15] That the Creature later feeds his creator in the Arctic, after receiving neither literal nor figurative sustenance from him, only reinforces Frankenstein's carelessness in the novel.[16]

Whether the Creature is *homo faber* is in some ways an unremarkable way of defining the human versus animals. Rather than affirming the validity of such distinctions, I argue that Mary Shelley demonstrates a cultural-technical understanding of the Creature's humanity in her multiple references to food and how it is produced. The Creature's technical deficiency is placed in a wider reproductive context in the sense that how the Creature is or is not fed is a measure of the human milieu that produced but does not sustain him. Food techniques belong to this reproductive milieu in an often obscured level of mediation.[17] Despite our enduring fascination with the novel's depiction of burgeoning fields of science, *Frankenstein* also includes technoscience in its depictions of food techniques, and Mary Shelley accounts for their profound powers of mediation.

To continue, let us turn briefly to the concept of cultural techniques from the Canadian-German media theory traditions. This concept includes the role of a variety of techniques in the reproduction of human social life, particularly in its state forms such as through filing systems.[18] As captured by the original agricultural sense of the word (*Kulturetechniken*), cultural techniques of food production constitute a fundamental part of social organization. Bernhard Siegert significantly argued in 2013 and 2015 that food techniques were the first cultural techniques and that they constituted the human as such.

In *The Raw and the Cooked*, Claude Lévi-Strauss argues that cooking and early forms of food processing are the original sites of distinction between nature and culture and between humans and other beings.[19] As we began to literally process our surroundings through ever more complex cooking or preservation techniques, we gave shape and language to a distinctly human culture and world. Indeed, our early food techniques are arguably *ontogenetic*, that is, preceding and constitutive of the human as such. Cooking released the human mouth from the exclusive acts of eating and breathing and, in the process, created one of the conditions for our capacity for language.[20] In effect, cooking predigests food.[21] As a result, early humans required less time for mastication and digestion, accelerating the release and intake of energy and nutrients, as well as making available previously inedible foods, too firm or toxic for the human digestive apparatus. Freed from this work, the facial-cranial architecture of the human transformed; encephalization—the evolutionary increase in brain size relative to body mass—is in humans arguably a result of this increased and accelerated access to nutrients that cooking enabled.[22] Lewis Mumford has gone as far to speculate that our prehistoric plant-gathering activities may have been the basis of language formation, that is, of our first vocabularies.[23] In the act of searching, tasting, and identifying, we learned to distinguish "insect from herb, and by degrees, one herb from another" (II:III.81), as the Creature does in *Frankenstein*. Along with our early food techniques, this knowledge allowed humans to become food generalists,[24] eating a wider variety of foods than other species and making our chances of survival greater.[25]

Bruno Latour argues in *We Have Never Been Modern* that "the very notion of culture is an artifact created by bracketing Nature off."[26] Likewise, the very notion of the human is an artifact created by bracketing off the animal. Both Lévi Strauss and Siegert cite animal husbandry as an important technique of hominization, particularly in relation to the animal pen or kennel.[27] Generally, the concept of cultural techniques describes how our *ontic* or material activities mediate and help to process distinctions about ourselves and the world.[28] "To investigate cultural techniques," writes Siegert, "is to shift the analytic gaze from ontological distinctions to the ontic operations that gave rise to the former in the first place."[29] For example, a door processes the distinction between inside and outside, a not-insignificant distinction for, say, a Black man outside a whites-only club in the 1950s United States or a prisoner inside a jail cell. Similarly, for Siegert, the pen or enclosure is a cultural technique that materially processes the distinction between human and animal.

Siegert draws on Johann Gottfried Herder's theory of language, which argues in contrast to Mumford's focus on plant gathering that "the end of foraging is the precondition for the acquisition of language."[30] For Herder, a naming system could not exist until the human was bracketed from nature by the animal pen: "It is the pen that keeps humans from attacking sheep in the manner of wolves and lions. . . . The sheep-man impelled by instinct to assail sheep is a monster, a humanimal, a being in which the difference between animal and human is suspended. It is the pen that first allows for a theoretically reflected relationship between man and sheep, an intellectual rather than instinctive relationship."[31] While the pen and animals within it are the ontic preconditions of an essentialized concept of the human, as soon as the human has been ontologized, the things and beings that participated in this bracketing are disavowed. As Siegert argues, the ontological distinctions between human and animal are slippery, if considering other animals' capacity for signification or sociality; the enclosed pen becomes necessary for maintaining the physical separation between humans and animals that our ontological distinctions cannot.[32] "The corral that excludes 'desire' or 'instinct' from sheep-shepherd relations serves to include animals in the phylogenetic origin of language. At the same time, it becomes necessary to devise cultural techniques aimed at excluding animals from the ontogenic origin of language" and the ontogenesis of the human as such.[33] The Creature is the subject of just such a technique; he is utterly divided from the cottagers by the animal pen.

Like the prepen "sheep-man," the Creature embodies that difference between animal and human. The fact that he forages and gathers acorns, historically associated either with famine or "uncivilized" peoples, codes him as less-than-human.[34] For example, one nineteenth-century writer describes acorns as "the fruit of the oak, and if they are planted oaks will grow from them. Uncivilized people often eat acorns, but as they have a bitter taste, civilized people, having other articles of food for themselves, leave the acorns to pigs, deer, squirrels, rats, mice, and birds."[35] The Creature's command of human language, inherent sense of morality, and empathy, however, seem to qualify him as human. He eventually masters the Promethean flame.[36] Following Siegert, however, even though the Creature acquires language, his nonhumanness is reinforced by the pen he inhabits. His efforts to transgress those boundaries ultimately fail, as do the Creature's attempts to commune with humans. Ontically he remains distinct from them, a distinction the Creature himself verbalizes in relation to food when he bargains with Frankenstein for a companion. He declares that his "food is not that of man" and he swears to leave human civilization with his companion and subsist on acorns and other wild foods (II:IX.120). Food was what first drove him toward human society, and through it, he also signifies his willingness to break with the human world.

With this analysis, my intent is not to reproduce an essentialist notion of the human by calling the Creature an "animal" but rather to point to the technical and social systems that mediate such distinctions. Despite the Creature's exclusion from humanity in the De Laceys' kennel, a boarded-up door suggests historically that the animal did not always remain so separate. Indeed, our food systems ontically affirm a hybridity of the human and animal. We drink another animal's milk, for example, calling into question how humans and animals are defined and how feeble those definitions can be. Our food systems periodically break down, as do the ontological distinctions they produce, for example, when communicable diseases like the swine or avian flus break the "species barrier." As Mary Douglas observes, when such "monstrous births occur, the defining lines between humans and animals may be threatened."[37] However, the "monstrous births" of any system are endogenous to them. Mark A. McCutcheon, citing Marshall McLuhan in his contribution to this volume, observes that all technology can be described "in terms of invasion, disease, disaster," terms that can indeed be extended to our most ancient and modern food techniques. For instance, it is estimated that over eight hundred of the fourteen hundred known human pathogenic organisms are *zoonoses*, viruses and parasites of animal origin introduced in the course of our progressive domestications and cohabitation with animals bred for food.[38]

Cultural techniques have the power to form and reproduce ontological distinctions, such as those distinguishing the human from animal. The Creature endures this in his continual exclusion from human society via their food techniques, and as a result, he is coded as animal or nonhuman. However, as Tero Karppi observes, cultural "techniques that draw distinctions are not static. . . . Distinctions are drawn and redrawn."[39] Given my argument that food techniques are ontogenetic of the human, it follows that if our food techniques change, so should the human; food holds a key to understanding the human and its limits, both ontologically and ontically. This is in large part because our need for food poses a biological limit and, at present, a terrestrial one to any human project interested in preserving (in time) or extending (in space) human life. Much of what we eat can be drawn back down to the ground, the *humus*, and much of how we conceive of the human can also be drawn back to it; Martin Lee Mueller argues that it is where we get our name.[40] Could we, then, become *posthuman* through new food techniques?

Posthuman Food for Inhuman Spaces

To explore how humans can become posthumans through the invention of new food techniques, I now turn to Tsutomu Nihei's *Knight of Sidonia*. This fifteen-volume manga is set in the year 3394—one thousand years after the unfathomable and near-indestructible alien race called Gauna destroys the Earth. Though hundreds of "seed ships" left Earth upon its destruction, only Sidonia appears to remain. Probing questions of mortality and transcendence in the context of a post-Earth civilization, *Knights of Sidonia* takes up several themes resonant with *Frankenstein*. There are mad scientists and laboratory-born alien-human hybrids; it is a cautionary tale about the dangers of science and technology run amok. But it is overall a story of survival, one revealing of the pivotal role that food techniques would play in any extraterrestrial journey for humans. Interstellar travel will rely on humans being able to reproduce their food techniques in alien or closed-system spaces like spaceships or else find ingenious ways to bypass the organic limits of our bodies and the bodies we eat. What is therefore particularly interesting about the Sidonians is that they are photosynthetic, allowing us to speculate on whether the boundaries of the human will one day be redrawn and whether food techniques will lie at the heart of the distinction. The analysis that follows also parallels *Frankenstein* and *Knights of Sidonia* on the basis that both narratives include journeys *post-humus*, to the sterile Arctic and the depths of outer space. Therefore, before I speculate on the possibility of humans becoming photosynthetic, I discuss the kinds of food techniques that made such expeditions possible and that are the progenitors of the food developed for space travel.

Food, though composed of organic materials, takes a distinctly historical and technical shape, owing to thousands of years of collective development of food techniques, from cooking and fermenting to cultivation and animal husbandry and to the more industrial food technologies we see today. Preservation and storage techniques (canning, salting, refrigeration, etc.) have allowed us to extend our sphere of activity into environments that are barren of food or at least providing much less of the foods we need. I refer to the sea, the Arctic, and outer space as *inhuman* spaces because they are inhospitable to the human, though they certainly can be *humanized* by way of technics. The question remains, however, just how far technics can take the human in these hostile places.

Our first travels *post-humus* occurred when we left land for the sea. The sea, as John Durham Peters observes, is not a home for humans as it is for the cetaceans and dolphins that see, speak, hear, and feel differently because of their aqueous nature.[41] Great human effort and ingenuity is required to set foot in this environment, which is why "life at sea" is constituted by the ship, itself a conglomeration of technologies and techniques that help transcend our terrestrial bounds. Referencing Buckminster Fuller, Peters calls the ship an "arch-medium" of humanity: "As the first completely artificial environment for human dwelling, the ship is an allegory for civilization. . . . The sea was where human invention . . . most decisively emerged, since the sea forced people to live by their wits and carried them around the world; thus people became *generalists*. The ship not only makes the sea available; it makes features about social order visible. On a ship *physis and techne are one*."[42] However, though food techniques do not presume a completely artificial environment, they nevertheless altered environments and produced the kinds of foods that provisioned and thus made sea travel viable. In the form of food, "physis and techne" are more primordially one, and as previously argued, food techniques are the first site of human generalism. A ship may indeed "[stand] in for nature, replicating what the terrestrial environment provides—footing, water, food, shelter, sleep, waste disposal—for an extended period on the open seas,"[43] but it cannot replicate cultural techniques of the soil and must rely on them for its "extraterrestrial" sustenance. We have yet to launch a ship that does not need to return to earth or Earth for provisions.

The business of developing foods that could travel long distances and be stored for long periods of time belongs historically to the realm of war and conquest (sometimes called "exploration") as well as laying the foundation for national and global logistics networks for the movement of food. Sue Shepherd observes that dried foods were "vital for both people and armies on the move since ancient times, establishing from the very beginning a pattern that sees the demands of the military shaping food preservation techniques right up to the modern day."[44] Nicolas Appert's early process for canning foods in glass jars, lauded for "fixing the seasons," was considered "a French military secret."[45] Tin-canning was popularized after proving essential to provisioning in the US Civil War.[46] Dehydrated, freeze-dried, and vacuum-packed foods, although common consumer products now, were engineered for the military and space travel.

Given the late eighteenth-century setting of *Frankenstein* and that Arctic exploration did not really take off successfully until later in the nineteenth century,[47] expeditions like those led by Captain James Cook (1768–1779) offer some insight into the kinds of provisions Walton's ship would carry. Canning and refrigeration had not yet been developed in this period (the French Navy began to trial Appert's canned foods in 1803),[48] and therefore we can assume that Walton's ship carried predominantly dried, salted, pickled, or otherwise "cooked" foods, for example, the "portable soup" that the British Navy began purchasing in 1756 for its ships.[49] In Cook's journal for July 1772, he gives the following account of the provisions placed aboard the *Resolution* and *Adventure*: "Biscuit, flour, salt beef, salt pork, beer, wine, spirit, [dried peas], wheat, oatmeal, butter, cheese, sugar, [olive oil], vinegar, suet, raisins, salt, malt, [sauerkraut], salted cabbage, portable broth."[50] Livestock, too, was carried on board.[51] Cook's mission to circumnavigate the globe, however, anticipated human contact and access to trade or wild food sources, something Walton could not expect given the sparse populations and even sparser flora and fauna of the Arctic tundra.

According to Charles E. Robinson, Mary Shelley added the Arctic framing narrative at a later stage in the novel's writing, suggesting that it plays a deliberate role in how we are to evaluate the actions of Frankenstein and Walton. Though Walton's journey would be impossible without food, he makes no reference to provisioning in his letters, unlike Cook in his journals. He communicates musings and aspirations for the trip solely as a journey of self-realization. According to Marilyn Butler, "Mary Shelley's approach to [Walton] is unmistakably sardonic. She portrays him as a spoilt young egotist who complains continuously . . . and all along risks the lives of the family men who make up his crew."[52] Indeed, we learn from the novel's first sentence that even dear Mrs. Saville regards her brother's enterprise with "evil forebodings" (I, Letter I.5). This is not surprising given the dissonance that Walton demonstrates in his perception of the Arctic and reports of its inhuman climate: "I try in vain to be persuaded that the pole is the seat of frost and desolation; it ever presents itself to my imagination as the region of beauty and delight. . . . The sun is forever visible; its broad disk just skirting the horizon, and diffusing a perpetual splendor. . . . Its productions and features may be without example, as the phænomena of the heavenly bodies undoubtedly are in those undiscovered solitudes" (I:Letter I.5). Neither stellar bodies nor the Arctic sun can feed humans without the mediation of soil, plant, and animal to convert their energy to food. Like Victor in his mad moments of creation, Walton thinks of nothing but adventure. For example, he remarks on his crew hunting as if for sport (I:Letter II.9), but surely, in the cold environs that place larger energy demands on the human body, these activities supplied necessary fat and calories for the men. What other trials the crew experiences are unnarrated, eclipsed especially when Victor is rescued and begins telling his tale to Walton.

After following the journeys of Victor and the Creature to their conclusion in the Arctic, we return to the framing narrative of the novel. The once-enthusiastic captain is surrounded by a crew who is on the brink of mutiny, for they have been led unprepared into this inhuman environment and they demand to return home. The captain can no longer guarantee the life of his crew. He seems as careless with life as Frankenstein with the Creature. Victor's last feverish oratory about the heights that human endeavor can reach, telling the sailors to "be more than men," leads Walton to deduce that he would "rather die, than return shamefully,—[his] purpose unfulfilled" (III:VII.183). And yet he does return, drawn back to the world that supplies his human needs. Over and over, the characters in *Frankenstein* are humbled by their relation to food, their mortality and necessity raised in relief no matter their relative distance from food techniques.

Knights of Sidonia poses more humbling prospects yet. As I discuss next, our relatively brief history of space travel attests that food in these environments must fit into a highly rationalized system that prioritizes nutritional over cultural or aesthetic value. The vision of humans traveling beyond Earth to the far reaches of space is one of transcendence but would mean great sacrifice, of bodily health and human culture, particularly in the form of food "cultures" that cannot be replicated in an artificial environment. As a closed system, the spaceship must be an arch-medium unlike any human ship now existing. If it is to itself provide food for its passengers, it must possess the media that transform inedible organic matter into food for humans. What qualifies as food for humans will be redefined under the circumstances; new cultures would emerge, to be sure, but whether they will be human cultures is another question. At the same time, and following Karppi's argument, the foods we have developed for long-distance travel seem to suggest that we were always already posthuman, particularly those foods that have fueled transoceanic and space travel thus far. After all, both sailors[53] and astronauts[54] return to the Earth physically different than when they left. Food-preservation techniques of the late nineteenth and twentieth centuries, along with industrially produced food surpluses, also allow billions of people to survive with little or no direct ties to the soil. In other words, food techniques are already processing a distinction between humans who are intimately tied to the soil and those who live a kind of *post-humus* life, a division of labor that many take for granted given that all humans need to eat. Hence, "eating the sun" as the Sidonians do would liberate the human from its terrestrial roots in a way that present food techniques cannot.

Eating the Sun

As Mary Roach explains in *Packing for Mars*, any spaceflight requires "carefully measured intakes and outputs"[55]—the entirely man-made environment of a spacecraft cannot tolerate unintended cargo or human error. Food poses a challenge for spaceflight. Our need for food makes it an absolute necessity. How to contain its organic and perishable qualities is a matter of concern not only for food preservation but also for preservation of complex machinery and circuitry that would be ruined by crumbs or liquids.[56] In the closed system of a ship, the necessity of excretion likewise cannot be overlooked; the abject wastes of the body must be named and accounted for.

According to Roach, space food technologists are concerned with making lightweight and compact food, for "every extra pound that NASA launches into space costs thousands of dollars in fuel needed to lift into orbit."[57] They seek to improve caloric density, "packing the most nutrition and energy into the smallest volume of food."[58] Roach adds that polar explorers "facing similar constraints and caloric demands . . . pack sticks of butter."[59] In the early days of the US space program, astronauts and livestock became unlikely analogs: foods in cube form, tubes, and pellet form were given to astronauts by scientists, including veterinarian scientists, who understood the act of feeding in only the most practical terms.[60] Clearly, the development of food for inhuman spaces required (at least initially) *de*humanizing food, as well as the astronauts destined to eat it. Not surprisingly, the cost and complexity around feeding astronauts and dealing with their waste had some people seeking solutions that would reduce food to the absolute minimum needed for spaceflight—whether that be food in pill form or sending astronauts with enough body mass that their bodies would feed off their own fat stores, instead of needing constant replenishment with food.[61]

Provisions will only take the human so far when traveling inhuman spaces. Space food takes up vital real estate on ships, and most rations only last a year before beginning to lose their nutritional value and what little palatability they have. Twenty-four months is the longest shelf life of any NASA-approved ration, falling short of the three to five years it is estimated a manned mission to Mars would take.[62] Besides, the "way space travel works today, it's almost impossible to take along all the resources you need. That is why we have to develop a biological system so astronauts can produce their own food, and recycle all of the resources," says Silje Wolff of the Centre for Interdisciplinary Research in Space.[63] We see such a system in *Knights of Sidonia*, for instance, in the form of organic conversion reactors that provide fertilizer for food cultivation, made from all the ship's waste, including its dead. Sidonia also houses rice factories and a whole level dedicated to aquaculture.

Growing food in space today has limitations because of the zero-gravity environment (Sidonia has the technology to replicate gravity), the lack of mineral resources and physical space, and of course, the unknowns of growing plants beyond the *humus*. So far, lettuce has been successfully grown and eaten on the International Space Station in 2015.[64] Further attempts to cultivate zinnia flowers, in anticipation of growing flowering plants like tomatoes, were thwarted by mold, though a few specimens survived. "Space gardening" is still very much in the realm of trial and error, whereas long-distance space travel requires certainty. Production also takes place on a small scale, nowhere near the needs to feed interstellar travelers or settlers. Despite the limitations, positive psychological and social effects are already evident to researchers who see space gardening as not just a necessary food technique but one that will humanize inhuman spaces. Continuing on with these space food experiments, "scientists can establish their variables and thus nail down the specific conditions that cultivate plants—and how those plants can cultivate humans."[65] Replicating cultural techniques in space is here seen as an assurance for the survival of something ontologically human.[66] Humans capable of photosynthesis represent the extreme opposite of this scenario and an arguably attractive solution to many of the limitations that food poses for space travel.

In *Knights of Sidonia*, the impetus giving rise to the photosynthetic human was a battle four hundred years before the narrative present, when 99 percent of the human population no longer exists. With the food-producing infrastructure in tatters and to replenish the ship population, the remaining Sidonians decide to manufacture and genetically modify a whole generation of humans with the ability to consume solar energy directly. The "new species" is in some ways more perfect than anything Frankenstein imagined—nearly invulnerable to hunger, the Sidonians transcend the need for human cultural techniques, though what they refer to as "cultures" are still produced on the ship for their weekly meals and as part of the broader culture of social eating and festivals. In addition, a new "food" culture emerges around the act of disrobing and basking in photosynthesis chambers, either with members of the same sex or as a form of intimacy, alone with a partner.

The Sidonians, ontically, in their genetic makeup and physiology, are different from terrestrial humans. The ability to photosynthesize, to functionally convert solar power to human energy, represents a form of transcendence that no food technique has given us before. The media of plant and animal that preserve and transform energy into food for humans are dispensed with entirely. *Humus* or soil would no longer be a tether. Instead of eating the food "of man," we could "eat the sun."[67] As unbelievable as it may seem, scientists are actively considering how we might develop food techniques from this scenario. For example, a 2014 study by Julie Schwartz, Nicholas E. Curtis, and Sidney K. Pierce reportedly unlocked the ability of a sea slug called *Elysia chlorotica* to consume algae and "pirate their photosynthesis genes"[68]—leading us to wonder if humans could do something similar. The answer, for now, is no. The Harvard geneticist Christina Agapakis says that the likelihood of a photosynthesizing human is close to nil: "With leaves, plants are able to harness an enormous amount of solar energy relative to their size. Thick, fleshy humans, with our low surface-area-to-volume ratios, probably don't have the necessary bandwidth."[69]

Some people would breathe a sigh of relief that this transcendent notion remains impossible for humans. However, human food techniques, particularly in their industrial agricultural forms, are sites of great social and economic inequality, environmental destruction, and the reduction of ecological complexity to only those plants and animals that feed us. Given the chance to modify ourselves and no longer rely on heretofore "human" food techniques for sustenance, are we not bound to consider alternatives? The existence of this study indicates that the fictional scenario I analyze here is, in some manner, being pursued today (however futile it may seem). We are, nevertheless, compelled to at least imagine the benefits and consequences of this fictional food technique, knowing full well that fiction has many times become reality. Would becoming photosynthetic liberate the human, or would it create further distinctions and classes of being, reinforcing rather than overcoming the inequalities of human food systems? For instance, the proliferation of photosynthetic beings on Sidonia makes those with more "primitive" physiologies less-than-human and places greater burdens on a system that is increasingly modified to serve the new human.

In *Knights of Sidonia*, the protagonist, Nagate Tanakaze, cannot photosynthesize and still needs to eat daily. Thus, much of the narrative in the early volumes of the manga highlights the social, physiological, and technical distinctions that mark him as different from the photosynthetic Sidonians. When we first meet Nagate, he is looking for rice. He has lived his whole life in the bowels of Sidonia with only his grandfather, who warns him never to stray from their immediate environs. His grandfather dies, however, and the last of their food runs out. Like Mary Shelley's Creature wandering, hungering toward the human food in nearby villages, Nagate finally ventures into once-prohibited areas of the ship. He is forced to confront the Sidonians when he gets sucked into the rice-packaging machinery and is delivered into the hands of his unknowing benefactors.[70]

Nagate is an outsider, not belonging to the social division of labor and unable to eat the food of the new photosynthetic human. He is initially spurned for stealing food, which is also the first human crime that Mary Shelley's Creature commits.[71] Comically, Nagate is criticized for his smell, a consequence of his diet and physiology differentiating him from the Sidonians; even the rumbling of his hungry stomach strikes them as alien. If, as I argue, food techniques play a defining role in the concept of species, from our twenty-first-century point of view, Nagate still very much resembles a terrestrial human, whereas the Sidonians could be considered posthuman. But from a Sidonian point of view, being able to photosynthesize is inherently human; Nagate cannot and is therefore ontologized, like the Creature in *Frankenstein*, as nonhuman (referred to by one Sidonian as a "demi-human").[72] In both texts, such distinctions are fluid in that they are formed and reproduced through food techniques, which themselves are historical and changing.

Conclusion

Our food systems affirm that, to paraphrase Latour, we have never been modern or human, dependent as we are on nonhuman being for our ontic and ontological existence. Conversely, humans distinguish themselves from animals through the systems we erected to supply our most basic need. Our food techniques ontically demonstrate the peculiar and profound ways our species reproduces itself. The so-called Anthropocene is an accumulation of distinctions humans produced as they mediated their relationship with their environment.[73] Some people, like Morton, argue that it is our first agricultural techniques that propelled us toward our present and unprecedented moment—the Neolithic revolution was, in fact, a devolution.[74] Regardless of the validity of such a claim, agriculture now is set to surpass fossil fuels as the leading driver of climate change.[75]

Like Mary Shelley and Tsutomu Nihei, I argue that we must put food at the center of our shared narrative, especially if we hope someday for a more sustainable, ecological, and equitable life for humans and the nonhuman beings that give us life. Food techniques mediate our relationships with friends and strangers alike, across dinner tables and global divides, and they are a constant touch point between humans and nonhuman nature. By placing food techniques first in our questionings about the future of the species, we must reckon with the way food reveals our vulnerabilities as humans, ontically and ontologically. This is perhaps why Mary Shelley supplies frequent remarks by the Creature about food: to put Victor's carelessness in greater relief. Frankenstein creates life without a thought for how to sustain it. We could describe our whole food system in similar terms. Do we humans have the capacity to put aside all we think is unique about ourselves and embrace our interdependence with nonhuman beings? In doing so, could we develop food techniques that exploit as little of our environment or fellow beings as possible? Putting a posthuman spin on food is a not just an interesting exercise but a necessary one if we are ever truly going to address our planetary crisis and the violence and inequities of our food system today. We need to ask what role food techniques will and shall play in the survival of the human, that is, *if* it should survive at all.

NOTES

1. Quotations follow the publication of 1818, as presented in *Frankenstein; or, The Modern Prometheus: The 1818 Text*, ed. Marilyn Butler (New York: Oxford University Press, 2008).

2. Zoe Sofia, "Container Technologies," *Hypatia* 15, no. 2 (2000): 181–201. See also Don Ihde's concept of "background relations" in *Technology and the Life World* (Bloomington: Indiana University Press, 1990), 108.

3. See, for example, Jean Retzinger, "Speculative Visions and Imaginary Meals," *Cultural Studies* 22, nos. 3–4 (2008): 369–390; Laurel Forster, "Futuristic Foodways: The Metaphorical Meaning of Food in Science Fiction," in *Reel Food: Essays on Food and Film*, ed. Anne L. Bower (Philadelphia: Taylor and Francis, 2004), 251–266.

4. Retzinger, "Speculative Visions and Imaginary Meals," 381.

5. Tsutomi Nihei, *Knights of Sidonia*, 15 vols. (New York: Vertical, 2013–2016).

6. Timothy Morton, "Frankenstein and Ecocriticism," in *The Cambridge Companion to Frankenstein*, ed. Andrew Smith (Cambridge: Cambridge University Press, 2016), 143–157. See also Lisbeth Chapin's contribution to this volume for further ecocritical analysis of the novel.

7. Morton, "Frankenstein and Ecocriticism,"148.

8. "Perhaps the pristine blankness of the Arctic landscape later in Frankenstein's narrative is an objective correlative for the blankness available at the form level in the guise of Mrs. Saville and whatever space she finds herself in." Morton, 151.

9. Using Martin Heidegger's concept of care (*Sorge*), Morton asks, "What are we to care for, for whom are we to care, what is care, how do we care, why care, who cares? . . . Perhaps the characters all care in all the wrong ways—too aggressively, too melancholically, too violently. Heidegger argues that even indifference is a form of care." Morton, 151.

10. Victor Frankenstein reflects on his relation to food only after he embarks on the lonely journey to create the Creature's mate, when the greater social nexus from which he draws sustenance is revealed. Settling on the most remote of the Orkney Islands, Victor remarks on its barrenness, which barely provides habitable conditions for its inhabitants, and he concludes that the island is suited for his work. The few inhabitants of the island are dependent on the mainland, and Victor on them, hunting and trading food and money for use of their cooking tools. In the final scenes of the novel, as Victor hunts his creation into ever more hostile environs, he is sustained by pitying peasants and, eventually, the Creature himself. See *Frankenstein*, III:II.136–137 and III:VII.174.

11. Morton, "Frankenstein and Ecocriticism," 153.

12. Paul Virilio tells us that what we know to be human, or any other ontological category, is a matter of proportion: "There is no object without proportions. There is no man without Dimensions. . . . To be a man is to measure basically between one and two meters. If a man measures 20 meters in height, it is not a man. This is unthinkable." Virilio, *Grey Ecology* (New York: Atropos, 2008), 27–28.

13. Lisa Nocks, "Frankenstein, in a Better Light," *Journal of Social and Evolutionary Systems* 20, no. 2 (1997): 138.

14. Jane Maienschein and Kate MacCord, "Changing Conceptions of Human Nature," in *Frankenstein, or, The Modern Prometheus: Annotated for Scientists, Engineers, and Creators of All Kinds*, ed. David H. Guston, Ed Finn, and Jason Scott Robert (Cambridge, MA: MIT Press, 2017), 215–221.

15. Maienschein and MacCord, 218.

16. Marilyn Butler argues in her introduction to *Frankenstein* that this moment signals the inversion of the parent-child relation embodied by Frankenstein and the Creature (see Butler, ix–li). Given the detail with which the Creature reflects on the availability or scarcity of food, his trouble learning food techniques, and his observations of the DeLaceys' poverty, the Creature understands more keenly than Victor ever does the role food plays in social reproduction. When the Creature leaves a dead hare for Victor during their Arctic chase, it is a moment of cruel irony, providing sustenance when he was given none.

17. See Sofia, "Container Technologies."

Zoe Sofia's feminist philosophy of technology informs this argument. Sofia, drawing on Lewis Mumford, argues that technologies that are not mobile and dynamic tend not to feature in accounts of technology. What she refers to as "container technologies" (those that preserve as they hold) also tend to be found in domestic and gendered spaces, putting in relief the way that patriarchal structures bias our understanding of technology.

18. Cornelia Vismann, *Files: Laws and Media Technology* (Stanford, CA: Stanford University Press: 2008).

19. Claude Lévi-Strauss, *The Raw and the Cooked* (London: Jonathan Cape, 1970).

20. John Durham Peters, *The Marvelous Clouds* (Chicago: University of Chicago Press, 2015), 80.

21. Kristin Gremillion, *Ancestral Appetites* (Cambridge: Cambridge University Press, 2011), 26.

22. Gremillion, 22, 17.

23. Lewis Mumford, *Technics and Human Civilization* (New York: Harcourt Brace Jovanovich, 1966), 81, 101.

24. Gremillion, *Ancestral Appetites*, 23.

25. Interestingly, for those who work in the field of artificial intelligence, the ability of humans to generalize from particular experience and apply that knowledge in other contexts is what makes human intelligence unique and difficult to replicate. See for instance Jerry Kaplan, *Artificial Intelligence: What Everyone Needs to Know* (New York: Oxford University Press, 2016), 5–6.

26. Bruno Latour, *We Have Never Been Modern* (Cambridge, MA: Harvard University Press, 1993), 104.

27. For example, Lévi-Strauss argues that a Tikopia myth describing the transformation of men into pigs probably "preserved the memory of a technique of hunting, . . . of driving [pigs] into enclosures where they were kept and fed before being killed according to need." Lévi-Strauss highlights how the enclosure or pig pen separates and differentiates one type of animal from another, creating distinctions where once there may have been none. Lévi-Strauss, *Raw and the Cooked*, 87; Bernhard Siegert, "Parlêtres: The Cultural Techniques of Anthropological Difference," in *Cultural Techniques: Grids, Filters, Doors, and Other Articulations of the Real* (New York: Fordham University Press, 2015), 61.

28. *Ontic* and *ontological* are Heideggerian categories, which I define rather simply: *ontic* refers to a thing's factual or real existence (in the Lacanian sense); whereas *ontologies* are essentially structures of meaning for the human (like Lacan's "symbolic"), which do not easily comprehend their ontic origin. Geoffrey Winthrop-Young says that "underneath our ontological distinctions (if not even our own evolution) are constitutive, media-dependent ontic operations that need to be teased out by means of techno-material deconstruction." Winthrop-Young, "Cultural Techniques and Sovereignty," *Theory, Culture, and Society* 30, no. 6 (2013): 12. We must, in other words, interrogate the techniques or objects that enforce the divisions and relationships that constitute human understanding, of itself and its world.

29. Siegert, "Parlêtres," 61.

30. Siegert, 61.

31. Siegert, 61–62.

32. Siegert, 61.

33. Siegert, 63.

34. Fernand Braudel, *The Structures of Everyday Life* (New York: Harper and Row, 1981), 112.

35. David Salmon, *Longman's Object Lessons: Hints on Preparing and Giving Them, with Full Notes of Complete Courses of Lessons on Elementary Science* (New York: Longmans, Green, 1895), 79.

36. By the time the Creature is reunited with Victor, he has apparently mastered this skill, lighting a fire by which he tells his short and tortured history. This detail, barely noticed by Victor, is later put in relief by the Creature's account of his efforts to produce fire. He might have mastered this skill through watching the De Laceys, but having been rejected by them as well, the Creature turns it on their cottage. He destroys every "vestige of cultivation" in the garden and burns the dwelling (II:VIII.113).

37. Mary Douglas, *Purity and Danger* (Binghamton, NY: Routledge and Kegan Paul, 1980), 39.

38. James C. Scott, *Against the Grain* (New Haven, CT: Yale University Press, 2017), 103, 100–101.

39. Tero Karppi, "Killer Robots as Cultural Techniques," *International Journal of Cultural Studies* 21, no. 2 (2016): 107–123.

40. "In the very name we give our species lies dormant the primordial experience of a deeply reciprocal participation between ourselves and the ground beneath our feet—between the human animal and the Earth." Martin Lee Mueller, *Being Salmon, Being Human* (White River Junction, VT: Chelsea Green, 2017), 89.

41. Peters, *Marvelous Clouds*, 63–65.

42. Peters, 104–105 (emphasis added).

43. Peters, 105.

44. Shepherd, *Pickled, Potted, and Canned* (New York: Simon and Schuster, 2001), 35.

45. Shepherd, 213, 238.

46. Shepherd, 246.

47. Butler, introduction to *Frankenstein*, xxxiv.

48. Shepherd, *Pickled*, 228.

49. Rachael Mamane, *Mastering Stock and Broths* (White River Junction, VT: Chelsea Green, 2017), 39.

50. Barbara Burkhardt, Barrie Andugs McLean, and Doris Kochanek, *Sailors and Sauerkraut: Excerpts from the Journals of Captain Cook's Expeditions All Pertaining to Food with Recipes to Match* (Sidney, BC: Grey's, 1978), 23.

51. Burkhardt, McLean, and Kochanek, 12.

52. Butler, introduction to *Frankenstein*, xxxv.

53. "Seafaring has a long history, but sailors, who work under challenging conditions during long sea-voyages, have increased disease mortality and morbidity compared with land-based workers." Weiwei Zheng, Ze Zhang, Cuihau Liu, et al., "Metagenomic Sequencing Reveals Altered Metabolic Pathways in the Oral Microbiota of Sailors during a Long Sea Voyage," *Scientific Reports* 5, no. 9131 (March 15, 2015): 1.

54. A study conducted at Colorado State University compared the health of the astronaut Scott Kelly after a year spent in orbit with that of his twin, finding "abnormalities such as inversions and translocations in some of Scott's chromosomes and some damage to his DNA, as well as changes in his gene expression. Beyond these genetic effects, Scott developed thickening in his retina and in his carotid artery. There were also shifts in Scott's gut microbiome that differed from those of his Earth-bound twin." Catherine Zuckerman, "One-of-a-Kind Study of Astronaut Twins Hints at Spaceflight's Health Effects," *National Geographic*, April 11, 2019.

55. Mary Roach, *Packing for Mars: The Curious Science of Life in the Void* (New York: Norton, 2010), 288.

56. Roach, 289.

57. Roach, 288.

58. Roach, 288.

59. Roach, 289.

60. Roach, 298.

61. Roach, 289–290.

62. Maya Cooper, Grace Douglas, and Michele Perchonok, "Developing the NASA Food System for Long-Duration Missions," *Journal of Food Science* 76, no. 2 (2011): R42.

63. Norwegian University of Science and Technology, "Growing Food in Space," press release, January 6, 2019.

64. Alan Yuhas, "Nasa Astronauts Take First Bites of Lettuce Grown in Space: 'Tastes Like Arugula,'" *Guardian*, August 10, 2015.

65. Yuhas.

66. Some effort has been made to humanize space food again. Astronauts today have more varied diets and even have the opportunity to cook, after a fashion; scientists on the International Space Station reportedly experiment using some of the two hundred different packaged foods provided on board and the station's food warmer to create new foods that are otherwise not on the menu. Take, for example, a "Christmas meal the astronauts prepared, remixing a variety of their core and bonus food items to arrive at a final menu of mesquite grilled albacore steaks, a dressed up cornbread stuffing, and crab salad." Ria Misra, "A Complete Guide to Cooking in Space," *Io9*, April 24, 2014.

67. Maddie Stone, "Eating the Sun: Can Humans Be Hacked to Do Photosynthesis?," *Motherboard*, February 10, 2015.

68. Julie Schwartz, Nicholas E. Curtis, and Sidney K. Pierce, "FISH Labeling Reveals a Horizontally Transferred Algal (*Vaucheria litorea*) Nuclear Gene on a Sea Slug (*Elysia chlorotica*) Chromosome," *Biological Bulletin* 227 (2014): 300–312.

69. Stone, "Eating the Sun."

70. Perhaps unsurprisingly for someone living on the margins and constantly struggling to find food, Nagate states his love for Sidonia's food first before its people: "From the day I was born I lived my whole life underground. . . . I got by the whole time with not much to eat. Maybe it's because of that, but I really love all the different foods, and organisms and all the people living here on Sidonia." Nihei, *Knights of Sidonia*, 4:142–143.

71. Unlike Shelley's Creature, however, Nagate is soon welcomed into Sidonian society. Nagate has also been raised by a loving father figure since birth, in a bid to overcome the socially stultifying effects of being lab-born like the majority of the genetically engineered Sidonians, who reach physical but not always mental maturity by five years of age.

72. Nihei, *Knights of Sidonia*, 1:140.

73. Geologists (and many scholars) are debating whether we have entered a new geologic age marked by the irreversible change that humans have made to the planet. According to the Nobel Prize–winning scientist Paul J. Crutzen, we have left the previous epoch, the Holocene, and entered the "Anthropocene." Various claims have been made as to when this epoch began, ranging from as late as the atomic age (post-1945) to as early as four to six thousand years ago, when large-scale agriculture began to take shape, thus reshaping the Earth. See, e.g., Robinson Meyer, "Geology's Timekeepers Are Feuding," *Atlantic*, July 20, 2018.

74. Timothy Morton, "She Stood in Tears among the Alien Corn: Thinking through Agrilogistics," *Diacritics* 41, no. 3 (2013): 90–113.

75. See, e.g., Josh Gabbatiss, "Meat and Dairy Companies to Surpass Oil Industry as World's Biggest Polluters, Report Finds," *Independent*, July 18, 2018.

CHAPTER 5

Reading *Frankenstein*'s Ecological Legacy

Lisbeth Chapin

Frankenstein began in the Alps. It was produced by cold and rainy evenings, as the literary luminaries involved were "crowded around a blazing fire" in the "majestic . . . environs of Geneva," one summer that was not a summer, thanks to the environmental impact of the Mount Tambora volcano the year before, in what was then the Dutch East Indies, now Indonesia. According to Percy Shelley's 1818 Preface to *Frankenstein*, after one evening of reading German ghost stories around the fire, "the weather . . . suddenly became serene," and they all left the damp indoors for "a journey among the Alps, and lost, in the magnificent scenes which they represent, all memory of their ghostly visions."[1] It could be argued, however, that one in this group, Mary Shelley, took her ghostly visions to the Alps, as she soon after sat down at her desk and began the real work of her story, conceived in "waking dreams" by the light of the moon—so she narrates in the Introduction to the 1831 edition (186). Mountainous topography, particularly in its frozen state invoking the sublime, is the novel's central ecological symbol, connecting the tundra of the North Pole, where Victor Frankenstein is discovered amid the ice floes by Robert Walton and his crew, to Mont Blanc. Victor's first approach to the summit of Montanvert after his housekeeper Justine's death, a scene that previously had filled him "with a sublime ecstasy that gave wings to the soul . . . to soar from the obscure world to light and joy" (so he tells Robert Walton), is soon eclipsed by his morose state of mind (II:II.70). The Creature traverses the hostile mountain to confront Victor, a place where he can assert his own physical superiority over his creator.

What do Victor and his Creature seek in this environment, and how are they changed by it? Mary Shelley's alpine world is more than a tourist site and sight; it is a glacial ecosystem—the grandeur of the mountain world subject to the environmental conditions that determine daily experience. At the same time, it is a psychological metaphor: both Victor and his Creature find solace there, and

Shelley situates the two to dramatize the antagonistic forces of life and death active within the meteorological conditions and ecosystem that framed the environment of her own inspiration.

Among *Frankenstein*'s many energies, it explores its readers' relationship with the world, particularly with the natural world and death, amid the conditions that stirred its story into being. Neither empirical nor theological in perspective, but drawing us over a cliff to a new sea, *Frankenstein* is so heavy with guilt, alienation, loneliness, grief, and remorse that readers are directed to their own internal navigation of this new creature coming toward them. The legacy of *Frankenstein* is its insistence that this horror of discovery also has elements of resolution: humankind's identity and function are inextricable from the natural world. Human responsibility involves our relationships to our own and other species, how we "read" each other and our differences. An ecocritical approach is vitally productive for this investigation.

Shelley's intervention is her creation of a unique organic being, and she meant this in closing the 1831 Introduction with the words, "I bid my hideous progeny go forth and prosper." (191). Among other afterlives, *Frankenstein* can prosper as an ecocritical investigation of social issues involving the messy parts of the Creature's assembly in Victor's laboratory, the moon's presence over water, and the mountains of ice where the Creature and Victor meet again. Ecocriticism of *Frankenstein* emerged in the 1990s, generally taking the Creature as its focal point. To Karl Kroeber, *Frankenstein* is "the first literary presentation of an ethical decision founded on the practical possibility of destruction of the entire human species." Observing that Victor Frankenstein practices science in the techniques of self-isolation developed since the Renaissance, he emphasizes the "scientific practice" of "intellectual experiment and speculation pursued entirely for their own sake."[2] This is also true of Victor's science in natural settings beyond his laboratory. Jonathan Bate's ecocriticism discusses care for the natural world;[3] and Timothy Morton turns the lens toward *Frankenstein*, which he calls

> an ecological novel precisely *not* because it compels us to care for a preexisting notion of nature, but because it questions the very idea of nature. Far from standing in for irreducible particularity—and hence ironically generalizing that very particularity—the Creature represents alienated generality. In the sense that his existence subtends our personhood, he figures forth an essentialist view of nature. But insofar as this nature is abject and its stitches are showing, this "essence" includes arbitrariness and supplementarity. The Creature is made out of any body, anybody. Frankenstein's Creature is not even an other because he cannot return our gaze or act as a blank screen. He is a horrific abject that speaks beautiful Enlightenment prose, a piece of butcher's meat with blinking eyes.[4]

Morton contends that the "augury of *Frankenstein* is the reverse of deep ecology": the "task becomes to love the disgusting, inert, and meaningless" because the "most ethical act is to love the other precisely in their artificiality, rather than seeking to prove their naturalness and authenticity."[5] *Frankenstein* reflects the concerns of Bate and Morton as it shifts from Robert Walton's story to Victor's story to the Creature's story and along the way involves Caroline Beaufort's story, Elizabeth's story, the De Laceys' story, and Safie's story, ending with Walton's story. Even Walton's ship, immured in the ice, cannot impede the movement of the narrative, despite his uncertainty about whether his letters to his sister, Margaret, will ever be received. *Frankenstein* ends in the ice world, the Creature disappearing into it and Walton having agreed to turn his ship around and head south.

Because every character's narrative in the novel is first person, readers encounter the expressive quality of description. The unexplained revelation of Victor's discovery of "the cause of generation and life" is his narrative climax: "from the midst of this darkness a sudden light broke in upon me—a light so brilliant and wondrous, yet so simple, that while I became dizzy with the immensity of the prospect which it illustrated, I was surprised that among so many men of genius, who had directed their inquiries towards the same science, that I alone should be reserved to discover so astonishing a secret" (I:III.33). Shelley shows us how one's revelations about the natural world expand inward as well as outward. Victor's enlightenment fails because his experience—as do those of the "men of genius" whom he emulates—lacks humanity and compassion. The ecosystems and animals around him may claim value of their own. Susan J. Wolfson remarks in her chapter for this volume that "in a smart guess at genetics, Mary Shelley decided to make Frankenstein's Creature an intuitive scientist"—and of a stripe different from his creator. Far from being a "mechanism shocked into robotic animation," the Creature "is a being of innate human curiosity, thrust into a natural laboratory."[6] Victor and the Creature work in parallel but antithetical laboratories, learning through experimentation, discovering the elements of life, demonstrating the results, traveling, reacting, and acquiring more experiences that spur the ragged progression or regression of their levels of self-awareness. *Frankenstein* confronts readers with the self-destructive ills of humanity in its ambition for the self-awareness to confront those ills.

Frankenstein locates humanity's investigations with four elements: fire, earth, air, and water. The novel's ecological events and scenes involve lightning, flora and fauna, vegetarianism, glaciers and mountains, graves and grave robbing, slaughterhouse materials, the weather, the moon, Walton's ship, Victor's boats, lakes, the sea, and the Orkney Islands, among others. Shelley develops the characters of Victor and the Creature in encounters with each of these and emphasizes important contrasts. One significant element is how they acquire and use language. Isaac Cowell argues that Mary (as well as Percy) Shelley is acutely attentive to the way that scientific belief operates within "a reflexive structure of feeling that proves unaccountable to others"—that may even depend "precisely on its lack of overt justification."[7] I would add to Cowell's argument that the empirical experience of nature for Victor and the Creature supplies justification. Nature speaks *with* both, in a novel vitally concerned with the relationship of beings (human and Creature) and environment, such as the hostile glacial world in the novel and at Mont Blanc, in the shadow of Mount Tambora, and specifically with the Creature's and Victor's relationship with death and the dead.

Earth: The Arctic

Because the Creature is a singular being, the symbolism of his character is more fluid than it might otherwise be. We can establish that symbolism as a mountain or glacial ecosystem or more dramatically ascribe it even to a geological event: Mount Tambora's volcanic eruption in 1815. Its destruction was immense in that region and across the globe for several years. Significantly, "scholars have identified Tambora as a key factor in the harvest failures and food scarcities across the globe" in the years immediately after, perhaps even figuring in the typhus and cholera epidemics of the period.[8] The Creature's most elaborate dialogue in the novel is with Victor on Mont Blanc, where the Creature seems an indigenous species, undaunted by the extreme cold and undeterred by the rugged terrain. Bill Phillips cites this scene to demonstrate that "it is as if the storm were responsible for the Creature's existence: it is 'illuminated' and 'discovered' by the lightning," noting that, "the weather, . . . which kept all indoors at the Villa Diodati in June 1816, led directly to the Creature's genesis."[9] Because the volcanic ash produced the extreme cold and rain during what has been called the "year without a summer" in Europe, prompting the dreary conditions for the writing contest from which the novel sprang, the Creature could be a symbol of the mountain itself. From a political perspective, the Mer de Glace is where the Creature's voice is as relevant (or irrelevant) as Victor's, in a conversation topographically above human society and its governments below. Influenced by C. F. Volney's *The Ruins* (one of his learning texts), the Creature is derisive of aristocracies, monarchies, and religion; he extends that derision to all human endeavors that presume to assert superiority and dominance over others. The Creature vocalizes that the mountain's own "ruins of empires" include not only the contemporary meteorological ravages of Tambora's volcano and the severe crop failures that resulted but also the starving mobs of citizens demanding aid from the local populations driving them away that were present within 150 miles of Mary Shelley and her circle as she began her novel that summer, Gillen D'Arcy Wood has discovered.[10]

The Arctic is one of the most notable settings in the novel, and its role in each narrative segment continues to expand. Andrew Griffin argues that "Walton's narrative, framing the novel, is entirely Arctic; Victor's and the Creature's stories are set much of the time amid the peaks and glaciers of Frankenstein's native Switzerland, sometimes in the almost equally desolate landscape of northern Scotland, and—finally—in the Arctic again, where the lines of all three narratives converge." The "land of mist and snow," as Walton calls it, quoting Coleridge, is "before our eyes from the beginning and, intermittently but with powerful effect, until the end."[11] Readers would have had regular news stories about the exploration of the Arctic against which to set *Frankenstein*. The first prospect for publication was John Murray, the publisher, as Adriana Craciun notes, "who held the legal rights to publishing all official Arctic exploration accounts and also published the *Quarterly Review*, but he had turned the manuscript down." Craciun clarifies the context of Arctic exploration at the time: "Readers of *Frankenstein* and of contemporary periodicals would have seen Walton's expedition not as a throwback to Coleridgean or Miltonic symbolism, but as a companion voyage to the two Admiralty voyages that in 1818 set off for the North Pole and North-West Passage. . . . Shelley had begun writing *Frankenstein* in 1816, and added Walton's Arctic frame around 1817, after she had begun reading the *Quarterly Review*'s accounts of Arctic exploration written by contributor and Second Secretary to the Admiralty John Barrow." Craciun admires Shelley's originality "in forging this new connection (between the long history of exploration and the recent history of literature) through the three male narrators' interleaved quests, through which we are invited to see Britain's Arctic fever in 1818 as part of the modern landscape of utopian thought." She also emphasizes that in "*The Spirit of the Age* (1825), William Hazlitt had connected the Admiralty voyages of Arctic discovery to Godwin's exploration of ideal reason in *Political Justice*," that is to say, "of the undesirability of reason 'no longer warmed by the affections' as a foundation for political justice."[12] *Frankenstein* may have initiated the "use of polar cold as a metaphor for emotional coldness," argues Francis Spufford: "Thanks to Hazlitt we can see that Shelley's brilliant, implicitly feminist critique of the denial of the 'domestic affections' in Godwinian radicalism was inseparable from the debates surrounding the high-profile but repeatedly unsuccessful attempts in Arctic discovery promoted by Britain's Tory government and press."[13] Jessica Richard first emphasized that "Shelley's novel, far from simply appropriating a topic of contemporary discussion uncritically, must be counted among these voices that censured the revival of British polar exploration," pointing out that "the much vaunted expedition of 1818 was a failure." To Richard, "Shelley's critique of polar exploration became even more overt in her 1831 revisions, where she heightens Walton's polar enthusiasm and expectations of glory and fame as an explorer," while conjuring "the fantastic visions that polar narratives licensed," also exposing the "uncertain authority of poetical description" on which Walton's—and Barrow's—polar projects depended."[14] More than ladylike reticence, Shelley's turn away from the claims of Romantic poetry is informed by her reading of science as a dangerously similar enterprise of creation. Walton's polar quest shows us the risks of a hubristic ethic of exploration, whether poetic or scientific, that irresponsibly creates "regions of beauty and delight" out of a world that is stark and cold.[15]

These contextual perspectives are verified in the activity of the Creature in the Arctic as Victor pursues him across the ice. The two characters are connected in a trajectory of only one direction—by the victuals that the Creature leaves for Victor and by the words (whole paragraphs) he inscribes, according to Victor, carved into rocks and trees. Even if Victor is delusional and therefore an unreliable narrator, at this point, the symbolism still holds: if the Creature symbolizes the Arctic or glacial ecosystem itself, his response to being conquered—to explorations of himself as the ecosystem—accentuates the imbalance of such an imperial ambition. In the novel, this manifests itself as Victor's arrogance, which cannot accept the Creature's vulnerability or the reflection of Victor's own as a human creature's reductive perspective of the ecosystems around him.

Nearly the entirety of the Creature's story occurs in a dramatic dialogue with Victor, at the ascent of a peak not far from Montanvert, in a "field of ice," as Victor describes it, while above him "rose Mont Blanc, in awful majesty" (II:II.71). When Victor implores and calls on the "wandering spirits" (71) to allow him a faint happiness—notably not invoking a Christian God—his incantation is answered by something akin to the spirit of the mountain itself. In fact, the Creature's function in the mountain's ecosystem is so imperative in the novel that it resounds mythologically. The Creature appears almost as Prometheus, when he approaches Victor for the first time on Mont Blanc: "I suddenly beheld the figure of a man, at some distance, advancing towards me with superhuman speed. He bounded over the crevices in the ice, among which I had walked with caution; his stature also, as he approached, seemed to exceed that of man" (72). If the Creature is superhuman, what is this Promethean, otherworldly figure bringing to Victor but his own lightning back to him? This confrontation of the myth destroys the altruistic hero of Prometheus that Victor sought to embody; the Creature explains to his creator and doppelgänger that the results of his discovery have established a relentless, suffering being that requires deep attention, personally and politically—in the most intimate unit of family as well as the most inclusive unit of a nation, not to be disentangled, according to Volney's theory.

The Creature not only learns the French language when the De Laceys teach Safie and, unbeknownst to them, the Creature, about revolutions and empires in the Comte de Volney's *Ruins of Empires* (1791), but he also must attempt "to capture geographically disparate histories in themselves and together," observes Ian Balfour.[16] The Creature, who is disoriented in his own self-identity and listening from his secret position in the De Lacey hovel, learns the French language from Volney's study of fallen empires and revolutions, a singular primer for a new student. Does the Creature speak to Victor in this voice of Volney-condemnations? "Yes, ignorance and cupidity! These are the twin sources of all the torments of man! Biased by these into false ideas of happiness, he has mistaken or broken the laws of nature in his own relation with external objects; and injuring his own existence, has violated individual morality; shutting through these his heart to compassion, and his mind to justice, he has injured and afflicted his equal, and violated social morality."[17] As Ann McWhir observes about this and the other books the Creature listens to and reads, including Milton's *Paradise Lost*, he "learns a mythology or ideology that must destroy him when accepted as truth: convinced that happiness is possible only in society and only through a loving relationship with his creator-God." He reasons that the "question of identity is a matter of ideology, and, unfortunately, the Creature either reads the wrong books or learns to read them badly."[18] Because Felix DeLacey is his only instructor, these ideas and ideologies also "construct" the Creature, as much as Frankenstein has his physical being, but the Creature is alone because his creator has broken Volney's "social morality" of including him in his family; in that way, nature has not broken such a bond. The Creature's origins align, if not with nations, with ecosystems; with terrain, if not with city-states.

It is significant that the Creature's place of meeting again with his creator is Mont Blanc, a site of fluid political identity. Nicholas Birns observes that in 1816, Mont Blanc was not on the border of France and Switzerland, as it is today, but "between Switzerland and the Savoy province of the Kingdom of Sardinia, . . . annexed by the French during the Napoleonic wars as a prize of conquest."[19] Mont Blanc, with its shifting national determinacy, is a native ecosystem for the Creature, who crosses boundaries that for him are meaningless; he is more a native of the mountain itself, unencumbered by humans competing for rights or national territory. *Frankenstein* is "concerned with the impossibility of stable dwelling," comments David Higgins; its "restless movements between different parts of Europe and the Arctic mimic the Creature's restless and ultimately fruitless search for a location where he will be able to dwell unmolested, as well as Victor's restless desire to shape his environment."[20] This search terminates for the Creature in the Arctic, as close as the Creature can come to a stable dwelling. And it is Victor's final home, where he is much more vulnerable than his Creature. Whereas "for Victor the glacier is a sublime backdrop to his anthropocentric imagination," Higgins concludes, "for the Creature it offers the possibility of a relationship with the world."[21]

Not only are the restless movements of the Creature, during which time Victor is oblivious to his Creature's fate after abandoned, those of a being invalidated by any family or government; his movements also symbolize the restless activity in graveyards and the corpses' vulnerability to grave robbers. The Creature's siblings are all the corpses once interred and now pulled from the earth and made into a spectacle on an anatomist's laboratory table; the Creature is the only one that moves across the earth, the body that rises now as an eight-foot giant to confront one of the violators.

Grave Robbing and Anatomy Schools

In *A Tale of Two Cities* (1859), Charles Dickens makes the graveyard of St. Pancras Old Church the very site for the thievery of Jerry Cruncher, a "resurrectionist," the common term for grave robbers, to "dig someone out of a grave" and sell the body to an anatomist.[22] St. Pancras Old Church was the original location of the grave of Mary Shelley's mother, Mary Wollstonecraft Godwin, a gravesite that the Shelleys visited often. While there is no mention of the Godwins or Shelleys in Dickens's novel, Cruncher imagines conversations with the dead that he unearths (and Victor Frankenstein has availed himself of a similar resource, the charnel houses). Mary Shelley's *Frankenstein* confronts established beliefs about exactly what purpose a graveyard serves; if it is to harbor and honor the dead, the Creature's existence refutes that assumption. In the novel, the Creature's primary function is to be an animated boundary of the living and the dead in order to challenge assumptions about the treatment of death and grief, graveyards, and the practices of the medical community itself.

Although anatomical dissection in Britain and Europe was practiced before the eighteenth century, Parliament's Murder Act (1752) made the medical dissection of all executed murderers compulsory, intended as a deterrent to the crime of murder, with this specter of offense to the corpse. But medical science was a stronger factor: with the demand by anatomy schools far exceeding the number of hanged felons, families who buried their loved ones feared for the corpses. On one occasion, in Lambeth, England, in 1795, a crowd stormed the parish graveyard only to discover many empty coffins; it was reported that "some, in a kind of phrenzy [*sic*], ran away with the coffins of their deceased relations."[23] However, if only the bodies of convicted murderers who had been hanged until dead could by law supply the anatomists of the day, then the character of Justine in *Frankenstein* is worthy of note. Justine is accused of strangling to death the child William, Victor's brother. Although innocent of the crime, she is convicted by circumstantial evidence; she maintains her innocence, even after being coerced into confessing by a priest, and is promptly executed. Mary Shelley's readers would know that if Justine had been hanged in England and her corpse subject to the Murder Act, her dead body could have been legally available as parts for Victor's female Creature; as it happens, similar laws for the acquisition of corpses were not in place in Scotland, where Victor begins to assemble the female Creature. Although "grave-robbing was rife in the Scotland that Mary Shelley knew as a girl" and Percy Shelley attended the lectures of the surgeon John Abernethy, who in 1819 proposed that pauper bodies be used for dissection,

David Ketterer finds no direct evidence connecting specific events or the controversy of grave robbing to any of the three editions of *Frankenstein* (1818, 1823, 1831).[24]

Although Justine's trial is set in Geneva, its event is one that the British reader could conflate with the Murder Act: that if the trial were in England, her corpse, as that of an executed murderer, would be dispatched to the dissecting table. This possible association aggravates further the offensive choice of Victor not to disclose his part in the murder. Victor's culpability in the crime, as the anatomist who created the Creature who murders William, is less an indictment of his own character than is his noting to himself, "Justine was a girl of merit. . . . Now all was to be obliterated in an ignominious grave; and I the cause!" (I:VII.57). He declines to relay any of this to the authorities for fear that his story would be considered "the ravings of a madman" (57), and his silence extends the disgrace of her grave to himself.

L. Adam Mekler asserts in his essay for this volume that Justine actually receives a fair trial but that her gender and socioeconomic status contribute to her conviction, especially regarding the responses of the spectators, who, as he explains, "even though they are not officials of the court, can be seen to exert some influence on the outcome of the trial, if not as much as the evidence itself, in their expression of indignation at Justine's perceived ingratitude" toward the wealthy and privileged Frankensteins. Mekler's comprehensive analysis includes historical parallels of the Revolutionary Tribunal in Paris and an uprising in Geneva in 1794, as well as the child custody case of *Shelley v. Westbrook*, which began during the writing of *Frankenstein*.[25]

The ignominy of Justine's death not only is an instance of an innocent person's being convicted of murder but also would call to mind for any reader in Britain or Europe the fact that her actual grave might not necessarily provide a deterrent to the "resurrectionists" or to anatomists, such as Victor, from defiling it. Mary Shelley presents readers with the complex convergence of death, grief, and the pretension of a scientific community that thinks it can overcome either.

The Creature and Vegetarianism

Human bodies were not the only materials Victor gathered for his humanoid Creature; he tells Robert Walton that both "the dissecting room" (of anatomists) and "the slaughter-house" furnished his materials (I:III.35). This combination of human and nonhuman parts, argues Stephanie Rowe, is yet another source of disruption in humankind's relationship with the natural world—the suffering of animals at human hands—in a narrative whose purpose seems to be to document the impact of grief and loss caused by this relationship.[26]

Like the Shelleys, the Creature is a vegetarian, and he explains this to Victor in an admonishment: "My food is not that of man; I do not destroy the lamb and the kid, to glut my appetite; acorns and berries afford me sufficient nourishment" (II:IX.112). According to Rowe, he represents the abused, voiceless animals of which he is partly composed, underscoring his choice not to eat animals or harm them. Rowe notes that "the slaughterhouse" (I:III.35) supplied Victor the remnants of cows, sheep, pigs, birds, and horses, and not just remnants from the meat market but also from vivisection and galvanism (though these animals were already dead). Victor "tortured the living animal to animate the lifeless clay" (35) and describes that his "limbs now tremble, and . . . eyes swim with remembrance of these tortures" (35).[27] Early on, Victor remarks, "my utmost wonder was engaged by some experiments on an air-pump, which I saw employed by a gentleman whom we were in the habit of visiting" (I:I.23). (The air pump was an apparatus that usually involved depriving birds of oxygen and then reviving them.) "*Frankenstein* takes its place in this debate," comments Rowe, "by voicing the perspective of animals through the plea of the excluded Creature, monstrously figured as 'anima,' for the just recognition of the irreducibility of his own existence and experience and, on that basis, for Frankenstein's benevolence."[28] There is no discounting the significance of the Creature's relationship with nonhuman animals in the novel, including those of which he is composed; he is an amalgam of rejection in the parts, human and otherwise, from which he is made. Assembled as he is from the debris and detritus of life, the Creature that Mary Shelley has constructed carries with him a record of all that he was, however unnamed or unidentified. He is an assembled ecological text.

Because the female Creature is never finished, is essentially a half-completed assembly of human parts—we have no particular details of the second experiment's materials otherwise—she is neither animal nor humanoid. Even so, she inspires in Victor a visceral terror from her potential as one who may conceive and bear offspring. As Anne K. Mellor proposes about the female Creature's role in the novel, Victor "is afraid of her reproductive powers, her capacity to generate an entire race of similar Creatures"; the fear is potent "female sexuality."[29] This is certainly true, especially in that Victor seeks to pursue Nature "to her hiding places" (I:III.35). Shelley's story warns readers that her society's dismissal and refusal to recognize women's sexuality as life-giving and worthy is dangerously blind, and such blindness is a source of cruelty that precipitates laws oppressive to women. Anatomists want, as Victor does, to root out the mystery of the generation of life, innate in female sexuality, in order to appropriate it because it confounds them, and if it has power, then that power should be theirs, as that argument goes.

"*Frankenstein* takes its mythic place as a cautionary tale about pride," argues Charles Robinson, most notably when "Victor warns Walton early in this narrative, 'Learn from me, if not by my precepts at least by my example, how dangerous is the acquirement of knowledge and how much happier that man is who believes his native town the world, than he who aspires to become greater than his nature will allow'" (I:III.33).[30] From this perspective, Mary Shelley indicates that scientists and anatomists are missing an attribute fundamental to a definition of motherhood: self-sacrifice. In contrast, Caroline, Justine, and Elizabeth each embody the version of self-sacrifice as inspired by the care of others, which Victor cannot comprehend, although he would defend it as such in his pursuit of the Creature. When Walton's crew demand that he turn the ship around, they are declaring that the sacrifice of others for the sake of scientific glory is morally wrong. The sailors can challenge Walton because of their ability to speak; the power of rhetorical persuasion and eloquence is put to the test many times over in *Frankenstein*, endorsing one of the most central invocations in the novel: "listen to me," "listen to my tale," "hear my tale" (I: Letter IV.16, II:II.73, II:II.74).

Victor and the Creature echo each other in these phrases throughout the narrative. Until the Creature learns to speak, he considers himself simply as another inhabitant of the forests and landscape through which he travels, and language is unnecessary. Victor reverses the responsibility of the creator-as-mother in sacrificing the being he created instead of sacrificing himself; at the very least, Victor neglects introducing him to language. When Victor refuses to satisfy the Creature's poignant plea for a companion, the Creature inflicts on Victor's family the fate that Victor has inflicted on him. Victor has a loving family but does not activate an intimate connection with them in his most desperate struggles; he refrains from telling them his tale and in that way is like an animal isolated within his emotional ecosystem.

In the Creature, Mary Shelley creates a character with no family, no medium through which to mute or mitigate the effects of his interactions with the world, cultural and otherwise. His character is formed by experience with nature and with humans. He sees animals as worthy of kindness, refusing to kill them for food. He sees humans as approachable until they attack and revile him, whereby he decides they are to be avoided. The De Laceys unwittingly offer the Creature the opportunity to observe a family unit, notably, at first without spoken language. Even as one only on the periphery of this setting, he insinuates himself emotionally into their family unit.

As impoverished exiles, the De Laceys, much like animals in the forest, offer the Creature further opportunity for study. The Creature "reads" them, before he even learns language. Alan Richardson identifies the Creature's abilities as something more immediately between language and interpretation, noting that the Creature's "silent observation of the De Laceys, who unknowingly harbor the Creature for some time, brings out his still more striking innate capacities for social cognition" and that the "Creature can accurately read emotions from facial expressions, though he has only had momentary contact with human beings up to this point: 'a young man met her, whose countenance expressed a deeper despondence'" (II:IV.85).[31]

The Creature's understanding of expression will allow him to apply this sympathetic understanding to the books from which he learns human language. This observation holds true and marks more exactly a turning point in the Creature's understanding of expression, information that he will apply to his understanding of language. From what he can deduce of the De Laceys, his anxiety about encountering humans begins to diminish, and he slowly enters the world of language in peace and some version of companionship; he then begins to create his own world of thought and reason, even more so when he finds the three seminal books—Milton's *Paradise Lost*, Plutarch's *Lives*, and Goethe's *The Sorrows of Young Werther*—as well as Victor's journal. He has been gradually transformed from an unidentified species in the ecosystem to a product of human society—ideas, behavior, reason, judgment, and prejudice—albeit without the direct love of "domestic affections" that can put one's life experiences into perspective and temper societal impact. His desire for female companionship to emulate the De Lacey family goes awry when he speaks to Victor, because Victor is still infused with his own sense of power as creator, is repulsed by the text of the Creature's misbegotten body, and filled with mistaken apprehensions about the Creature's desire for a companion. In this anxiety about a reproductive agenda, Claire Tuite relates the female Creature to Thomas Malthus's *Essay on the Principle of Population* (1798), a fear of "reproductive excess: never merely herself, she is always the horror of more mouths to feed, the motor of the 'geometrical ratio' of 'doubling population.'"[32] Considering the context of the starving mobs wandering throughout the countrysides of Europe that summer and fall of 1816, including in Switzerland while Shelley was writing *Frankenstein*, Gillen D'Arcy Wood has drawn attention to the blurred distinctions between thousands of poor and famished humans, an angry and vengeful humanoid, and wild animals in a forest.[33] In the Creature and his mate, Shelley dramatizes the awful danger not of humanoid Creatures propagating but of human beings propagating, human beings corrupted by pride and cruelty, perhaps desperate with rage and hunger. "The Creature and Victor have each killed the other's female companion and the hope of a future family," comments Ron Broglio; "they have entangled themselves in a different future, a pact of hate and revenge outside of community and the hope a future community would bring."[34]

The first ecosystems that the Creature encounters, "the forest near Ingolstadt" (II:III.76), as well as the De Lacey home he discovers, introduce him to family systems—not without conflict and hardship but coherent and predictable to him, once he observes them long enough. His first pleasure in becoming conscious of his environment is gazing at the moon: "the only object that I could distinguish was the bright moon, and I fixed my eyes on that with pleasure" (76). The Creature is as an infant in a world of confusing sensations, focusing on the moon as on the face of a mother. Similarly, he is soon after enwombed in the hovel, through which he observes the De Laceys unseen. When he finds a fire "left by some wandering beggars" in the woods alone, he is delighted at its application in both heating and cooking the nuts and roots he eats (76–77). In watching the De Laceys, he observes wonderingly the use of fire not only for heating and cooking but also for light by the use of candles. The De Lacey home is the perfect nexus for a Creature struggling to find his place in an ecosystem and in a society; the connectors begin with light, fire, and a sense of enclosed security.

Air, Fire, and the Nature of Home

As the Creature explains to his negligent creator, "the bleak skies are kinder" to him than are people; he delights in birdsong and in the "radiant form" of the moon (II:III.78, 80). "The barbarity of man" is worse than any winter season; he is alluding to one of La Fontaine's popular fables (1668–1694) in admonishing those who abuse the "gentle ass" with "blows and execration," much as the Creature is beaten by those whom he encounters (II:IV.83, 86). Unsurprisingly, the Creature prefers the environment around him: "my nocturnal rambles were an extreme pleasure" (II:V.89). By day, he is content to learn, and by night, he is content to be a creature of the woods. And so, in rejection by the De Laceys, he gives in to "fearful howlings": "I was like a wild beast"; and when he is done with bestial venting, he finds solace in nature: "the pleasant sunshine, and the pure air of day, restored me to some degree of tranquility," before he decides to attempt another encounter with the family (II:VIII.104). After he overhears Felix explain to the landlord that an unexplained horror has compelled the family to leave, the Creature aligns with a "fierce wind" that tears "along like a mighty avalanche" to stir "a kind of insanity in [his] spirits, that burst all bounds of reason and reflection" (106). The merging of these three elements—Creature, fire, and wind—destroys the De Lacey cottage, along with his hope of ever being admitted into their social unit.

As the Creature travels north to Geneva, he endures the harsh winter conditions yet is most fearful of encountering "the visage of a human being" (II:VIII.107). Spring and "the loveliness of its sunshine and the balminess of the air" inspire in him "emotions of gentleness and pleasure, that had long appeared dead" (107). It being too early for flowers in the environs of Geneva, his comfort is sunlight and heat alone. "Half surprised by the novelty of these sensations, I allowed myself to be borne away by them; and, forgetting my solitude and deformity, dared to be happy" (107). The first human being he encounters is the young girl who slips and falls into the stream, whom he rushes to save, only to find a "rustic," who tears the girl out of his arms and then shoots him. The Creature suffers as much from the sense of injustice as from the bullet lodged in his shoulder. From this point on, the natural environment has no more positive influence;

his sufferings are "no longer to be alleviated by the bright sun or gentle breezes of spring" (108). Immediately after this, he murders William and places the incriminating evidence of the miniature in Justine's pocket.

Here, Victor's sympathy is undone. As the Creature gazes on the dead boy, he tells Victor, his heart "swelled with exultation and hellish triumph"; he clapped his hands in glee and then frames Justine, by placing a locket in her pocket. If this seems fantastically improbable, so is Victor's response: that the Creature's "words had a strange effect" on him and that he "compassionated" the Creature enough to consider making a female Creature, instead of having an enraged response at facing his brother's confessed murderer (II:IX.113). This conversation has a fantastical quality. Victor and the Creature have created their own postlapsarian garden, of a sort, and the content of their speech and the manner in which they deliver it rises into a realm we are expected to enter, perhaps unwillingly, in order to arrive at some new meaning from this encounter.

This conversation on Mont Blanc between Victor and the Creature has more impact as a mythological scene than as a development in the plot, and the fire between them is an important symbol. In the introduction to *The Original Frankenstein*, Robinson identifies significant differences between Shelley's 1816–1817 draft and the published 1818 edition, after Percy added four to five thousand words in total to the novel. The original structure was as a two-volume novel, fifteen chapters in Volume I and eighteen in Volume II, with this second one "beginning dramatically with the Creature's narrative, as he and Frankenstein sit before the fire that symbolizes Promethean knowledge."[35] This will become the third chapter in Volume III of the three-volume 1818 edition. As Robinson sees it, Shelley has "subsumed and conflated the three basic Western myths about those dangerous consequences" of "the pursuit and the expression of knowledge": the narrative of Adam and Eve (and God and Satan), as emphasized by the Creature's reading of Milton's *Paradise Lost*, which focuses on the Tree of Knowledge and the sin of pride; the Prometheus tale of hubris, as introduced by the novel's subtitle *The Modern Prometheus* and by the many references in the novel to fire and lightning and sparks (recalling the fire of knowledge that Prometheus stole from Zeus to give to primal man); and Aristophanes's story in Plato's *Symposium* about primal man reuniting with one's second self after being cut in half for presuming to scale the heights of heaven to challenge the gods.[36] In the Creature's origin, fire sparks the creation of a new species; in that way, Victor's organic chemistry experiment is a success. However, it is the response to this new knowledge, that is, the lack of self-awareness that no scientific pursuit without the temperance of compassion can justify arrogance and pride, that determines this experiment finally as a human, not a humanoid, failure.

The Creature exists continuously as a nonhuman animal in every ecosystem, tolerating a variety of seasons and terrain, which he anthropomorphizes as companions, cruel or kind; he is not recognized as human and declares himself to be, when in distress, like a wild beast. In arguing with Victor about making a female Creature as his companion, he declares, "man will not associate with me" and "my companion must be of the same species" (II:VIII.110). Victor declares his ambitions: "A new species would bless me as its creator and source; many happy and excellent natures would owe their being to me" (I:III.34). Ashton Nichols remarks that in Charles Robinson's facsimile edition of the *Frankenstein* 1818 draft manuscript, the word "species" is revealed as Mary Shelley's third choice, after first considering "creation" and then "existence."[37] Nichols explains that "Mary Shelley did not know precisely what a 'species' was when she published the term in the first edition in 1818, but neither did the Comte de Buffon, Erasmus Darwin, or Charles Darwin," considering Buffon's definition of species as fixed and the Darwins' as exploring beyond that into the discoveries of evolving species.[38] Erasmus Darwin is named in the first sentence of Percy Shelley's Preface to *Frankenstein* as one who would have "supposed" the story "as not of impossible occurrence," and Mary Shelley might have known Darwin in her girlhood, as a friend of her father's.[39] Darwin's speculation about "spontaneous vital production" in *The Temple of Nature* (1803) might incline readers to consider Mary Shelley's story as a response to that theory, that the Creature is another kind of "spontaneous vital production," the first and only of its kind.[40]

The element of fire appears in scenes throughout the novel, as Robinson notes, and these are generally dramatic in the narrative. When Victor was fifteen, as he relays to Walton, he witnessed "a stream of fire" as "the dazzling light" of lightning when it reduced an oak tree to "a blasted stump," which prompts his interest in electricity thereafter (I:I.24). Victor describes his epiphany in the laboratory similarly as "a light so brilliant and wondrous" that he transfers it into his creation, to "infuse a spark of being" and perhaps revitalize Victor's own social being in the world (I:III.2, I:IV.37), having abandoned all contact with his family for two years at Ingolstadt. A flash of lightning on Mont Blanc illuminates "the wretch, the filthy daemon," whom Victor reframes at that moment: "the light of my own vampire, my own spirit let loose from the grave, and forced to destroy all that was dear to me" (I:VI.54). Victor defines lightning and electricity as that knowledge which has destroyed him, though sparking the Creature into life; to Walton, he declares, "I am a blasted tree; the bolt has entered my soul" (III:II.125). His doppelgänger, the Creature, enacts this destruction also by an intentional act: the conflagration of the De Lacey cottage is the Creature's destruction of the family unit he yearns to join, ridiculing the fire of candlelight and warmth into burning ash as of a funeral pyre. The Promethean allusions enhance the symbol of fire as a gendered masculine power that charges a male Creature, assembled by a male scientist, into being; alternatively, the moon as a gendered female power, in both Victor's and the Creature's perspectives, presents a source of light and knowledge with an impact equally as mythic in the narrative.

Water, Moon, and Tides

If the moon is a motherly face, calming the Creature, it often does the same for Victor: "all nature reposed under the eye of the quiet moon," Victor declares, during his miserable days on the Orkney Islands, and as the tides are in its pull, so too can the moon just as easily distress him when he sees "by the light of the moon, the daemon at the casement" (III:III.130). In fact, the moon and water align in *Frankenstein* at key psychological moments for both characters.

It is most often to bodies of water that Victor retreats for comfort and calm when he is agitated, but the waters beyond Geneva's lakes do not always comply. When he and his family return home after their final visit with the condemned Justine, Victor takes a boat onto the lake and passes "many hours upon the water" (II:I.66). He is lulled into a state in between his agonizing life on land and thoughts of death on the waves, so much so that he confesses, "I was tempted to plunge into the silent lake, that the waters might close over me and my calamities for ever" (66). Victor alone in a boat on the water is enwombed by it, seeking comfort and an escape from a world in which his mother has died and his grief for her and others remains unattended, a world in which he has rejected his family's gestures of emotional intimacy by remaining isolated in his own secret narrative.

The Creature, too, is isolated in his own secret narrative, but it is water that first reflects his own image back to him, when he, as does the god Narcissus, gazes at himself under the shadowy moon: "but how was I terrified, when I viewed myself in a transparent pool!" (II:IV.85). To one who defines the De Laceys as his "beloved cottagers" (II:VI.96) and describes the fictional Werther as "a divine being" (II:VII.97), the shocking image of himself as hideously different from the "cottagers—their grace, beauty, and delicate complexions" (II:IV.84)—gives him an intimacy with the water and the natural elements around him that accept his presence, in spite of his appearance. In *Frankenstein*, beautiful humans are good, and unsightly humans—rough or unattractive in appearance or demeanor—are not. The Creature, once harmless and good, decides in his ugliness that he must overcome an aesthetic barrier he accepts as indisputable, if he wants to find companionship with humans at all. His conversation with the blind De Lacey father is successful enough, until Felix walks in and discovers him; the father is an ideal human, as one who is kind, accepting, and nonjudgmental, who acknowledges the Creature's presence without rejection, replicating the Creature's experience in an ecosystem without humans.

When Victor travels with Henry Clerval to England and then alone to Scotland, the contrast to the Orkney Islands is purposefully stark, arriving at his new laboratory in a place that blurs the human-animal boundary. Walking through Windsor, "rambling in its beautiful forest," he remarks also on "the lovely Isis," the tributary of the Thames that flows through Oxford (III:II.124); these scenes provide an environmental pleasantness, which is the last that Victor experiences, dramatically different from what he faces in the Orkney Islands to prepare for the female Creature. He denotes the remote location in Scotland as "hardly more than a rock, whose high sides were continually beaten upon by the waves"; the soil offers little "pasture for a few miserable cows" of "gaunt and scraggy limbs," alluding perhaps to those limbs he would begin to dismember to make the female Creature as the butcher would do to the cows (127). He continues,

> On the whole island there were but three miserable huts, and one of these was vacant when I arrived. This I hired. It contained but two rooms, and these exhibited all the squalidness of the most miserable penury. The thatch had fallen in, the walls were unplastered, and the door was off its hinges. I ordered it to be repaired, bought some furniture, and took possession; an incident which would, doubtless have occasioned some surprise, had not all the senses of the cottagers been benumbed by want and squalid poverty. As it was, I lived ungazed at and unmolested, hardly thanked for the pittance of food and clothes which I gave; so much does suffering blunt even the coarsest sensations of men. (128)

Essentially, Victor in the Orkneys is living much as the Creature did in the woods, except that Victor does not consider himself in symbiosis with the community; his alienation parallels the Creature's experience of living in or near human society. In this strange outpost of civilization, Victor is as horrified at the Creature's image in the window as the Creature is of his own in the pool. Soon after, the female Creature project is abandoned, another confrontation flares up between Victor and the Creature, and the Creature rows away in a boat. The water becomes that which buffers and accepts human failings, from suicidal thoughts to the pieces of the "half-finished Creature," which look to Victor as if he had "mangled the living flesh of a human being" (III:III.133). Victor sails out under cover of night for his task, and at the moment that the clouds obscure the moonlight, he casts his basket of refuse into the sea, listening to a human-like "gurgling sound" as it sinks (134). However, this remote island location, "hardly more than a rock" (III.II.127), is not the one on which Victor, like Prometheus, will be condemned; the possibility for each character's redemption exists still as the narrative's effects continue.

The element of water merging with the moon, mothers, and mothering often but not always rescues Victor. In any case, he must rely on nature to mother him, now that he has rejected the attempt to create a female himself. Once he has unburdened himself of the debris of the half-articulated female Creature from his laboratory, he lies in the bottom of the boat, refreshed by the breeze that fills him with "agreeable sensations"; he hears "only the sound of the boat . . . through the waves," and the murmur soothes him like a lullaby into a sound sleep (III:III.134). Again, he is enwombed in his solitary boat, on the sea, in the dark. When the storm rises, he looks on the heavens, now covered by clouds, and finds no moon to light his way. He looks "upon the sea" now to be his grave, and, as happens more often than not when Victor is in dire circumstances, he faints (134). He succumbs to unconsciousness, reverting to that of an infant in utero; it is as if the waters overtake his consciousness in order to preserve not only his life but also his psyche, such as it is. Later, describing his stay in the Irish prison, he tells Walton, "my life appeared to me as a dream; I sometimes doubted if indeed it were all true, for it never presented itself to my mind with the force of reality" (III:IV.139). On his brief honeymoon with Elizabeth, Victor enjoys their boat ride with the soft breezes rippling the water, and they arrive at a "lovely scene of waters, woods, and mountains" to rest for the night (III:VI.153). Soon after, the clouds sweep across the heavens "swifter than the flight of the vulture" (153), dimming the moon, and a storm moves in.

The natural elements assume a similar emotive significance in these scenes as they do for the Creature in his first weeks of life, whose story Victor has unconsciously absorbed and incorporated into his own storytelling; the water, wind, and moon are enfolded as part of the intimacy of the narrative, and their influence on the character is part of our understanding of the story's impact. Once the "pale yellow light of the moon" (154) reveals the dead Elizabeth, akin to Victor's dream the night the Creature comes to life, and he sees the Creature's jeering face, he must pursue that which contains his deepest narrative, the Creature, because there is no family with whom to resurrect any human intimacy; Victor's self-righteous pride remains intact, and the only being with whom he is emotionally invested is the humanoid whom he pursues, whose own self-righteousness he fails to recognize.

Conclusion

Frankenstein may plausibly have been initiated by the fire of the Tambora eruption, causing the summer that was not a summer and a sun that could not shine through the ash. Shelley evokes this event at the end of the novel, closing the Creature's narrative with his declaration of a funeral pyre, leaving readers with the image of fire on ice, a volcano against the frozen earth. His "ashes will be swept into the sea by the winds," he eloquently describes to Walton, and in the last words of *Frankenstein*, the Creature is "soon borne away by the waves, and lost in darkness and distance" (III:VII.179). No part of the story is without an ecological setting that infiltrates the perceptions of the characters and determines our interpretation of the narrative, whether the character in that narrative is the moon, a humanoid, a child, or a storm. We do not have Walton's reflections on this story after the Creature departs, but we do know that Walton's relationship with his sister, Margaret Walton Saville, whose initials recall Mary Wollstonecraft Shelley's, as Martin Tropp (for M.S.) and then Anne K. Mellor (for M.W.S.) have observed, is one of writer to reader, in loving concern for her and for the manner in which she reads and responds to his story.[41] We assume that Walton's ship will move forward out of the ice, that he will find his way home; we have an affection for the lonely, letter-writing captain, the only human being with whom the Creature has an unmediated conversation. Margaret is privy to all three of the males' stories, and her silent listening is the first act of the novel's readership; through her brother, she reads about Arctic exploration, the terror and fascination of scientific discovery, and the tragedies that can ensue. In every significant scene of *Frankenstein*, the elements of nature are integral—as particular phenomena and as the material of myth.

Mary Shelley creates this story so that we readers, too, will move across the earth, through nations and ecosystems, encountering all of the life around us, expanding and extending a greater self-awareness of ourselves individually and acknowledging a reverence for the complexity of the earth and our intimacy with the entirety of it. As she understood, there is nowhere else to go, no matter the darkness or distance.

NOTES

1. Preface to *Frankenstein; or, The Modern Prometheus*. Quotations from the novel follow the publication of 1818, as presented in *Frankenstein: A Longman Cultural Edition*, ed. Susan J. Wolfson (2nd ed. New York: Longman, 2007), 5.

2. Karl Kroeber, *Ecological Literary Criticism: Romantic Imagining and the Biology of Mind* (New York: Columbia University Press, 1994), 28, 29.

3. See Jonathan Bate, *Romantic Ecology: Wordsworth and the Environmental Tradition* (London: Routledge, 1991); and Bate, *The Song of the Earth* (London: Picador, 2000).

4. Timothy Morton, *Ecology without Nature: Rethinking Environmental Aesthetics* (Cambridge, MA: Harvard University Press, 2007), 194.

5. Morton, 195.

6. See page 9 in chapter 1.

7. Isaac Cowell, "The Shelleys' Secret Convictions," *Essays in Romanticism* 24, no. 2 (2017): 146.

8. David Higgins, *British Romanticism, Climate Change, and the Anthropocene: Writing Tambora* (Cham, Switzerland: Palgrave/Springer Nature, 2017), 2.

9. Bill Phillips, "*Frankenstein* and Mary Shelley's 'Wet, Ungenial Summer,'" *Atlantis* 28, no. 2 (2006): 63.

10. Gillen D'Arcy Wood, "*Frankenstein*, the Baroness, and the Climate Refugees of 1816," *Wordsworth Circle* 48, no. 1 (Winter 2017): 3.

11. Andrew Griffin, "Fire and Ice in *Frankenstein*," in *The Endurance of Frankenstein: Essays on Mary Shelley's Novel*, ed. George Levine and U. C. Knoepflmacher (Berkeley: University of California Press, 1979), 54.

12. Adriana Craciun, "*Frankenstein*'s Politics," in *The Cambridge Companion to Frankenstein*, ed. Andrew Smith (Cambridge: Cambridge University Press, 2016), 91–93.

13. Francis Spufford, *I May Be Some Time: Ice and the English Imagination* (London: Faber and Faber, 1996), 87.

14. Jessica Richard, "'A Paradise of My Own Creation': *Frankenstein* and the Improbable Romance of Polar Exploration," *Nineteenth-Century Contexts* 25, no. 4 (2003): 307.

15. Richard, 308.

16. Ian Balfour, "Allegories of Origins: *Frankenstein* after the Enlightenment," *Studies in English Literature* 56, no. 4 (Autumn 2016): 783.

17. Constantin François de Chasseboeuf, Comte de Volney, *The Ruins; or, Meditation on the Revolutions of Empires and the Law of Nature* (1830), chap. 8.

18. Anne McWhir, "Teaching the Monster to Read: Mary Shelley, Education and *Frankenstein*," in *The Educational Legacy of Romanticism*, ed. John Willinksy (Waterloo, ON: Wilfrid Laurier University Press, 1990), 53.

19. Nicholas Birns, "Secrets of the Birth of Time: The Rhetoric of Cultural Origins in *Alastor* and 'Mont Blanc,'" *Studies in Romanticism* 32, no. 3 (Fall 1993): 361.

20. Higgins, *British Romanticism*, 68.

21. Higgins, 71.

22. Charles Dickens, *A Tale of Two Cities* (London: Chapman and Hall, 1859), book 1, chap. 3, "The Night Shadows," https://www.gutenberg.org.

23. Ruth Richardson, "Bodily Theft Past and Present: A Tale of Two Sermons," supplement, *Lancet* 364 (December 18, 2004): 44.

24. David Ketterer, "'Furnished . . . Materials': The Surgical Anatomy Context of *Frankenstein*," review of *Murdering to Dissect: Grave-Robbing, Frankenstein and the Anatomy Literature*, by Tim Marshall, *Science Fiction Studies* 24, no. 1 (March 1997): 121. See Marshall's book (Manchester University Press/St. Martin's Press, 1995) for an insightful analysis of historical background information pertaining to the conditions of grave robbing in the late eighteenth and early nineteenth centuries.

25. See Mekler's chapter for the extensive scholarly studies on Justine's court case, among the other trials in *Frankenstein*; for more on *Shelley v. Westbrook*, see also Lisbeth Chapin, "Shelley as Infidel: Child Custody, Poetry, and the Law," *Philological Review* 40, no.1 (Spring 2014): 57–84.

26. Stephanie Rowe, "'Listen to Me': *Frankenstein* as an Appeal to Mercy and Justice, on Behalf of the Persecuted Animals," in *Humans and Other Animals in Eighteenth-Century British Culture: Representation, Hybridity, Ethics*, ed. Frank Palmeri (Burlington, VT: Ashgate, 2006), 137.

27. See also Rowe, 137.

28. See also Rowe, 139.

29. Anne K. Mellor, "Possessing Nature: The Female in *Frankenstein*," in *Romanticism and Feminism*, ed. Anne K. Mellor (Bloomington: Indiana University Press, 1988), 220–232.

30. Charles E. Robinson, introduction to *The Original Frankenstein*, ed. Charles E. Robinson (Oxford: Oxford University Press, 2008), 31.

31. Alan Richardson, *British Romanticism and the Science of Mind* (Cambridge: Cambridge University Press, 2001), 161.

32. Clara Tuite, "Frankenstein's Monster and Malthus' 'Jaundiced Eye': Population, Body Politics, and the Monstrous Sublime," *Eighteenth-Century Life* 22, no. 1 (1998): 148.

33. Wood, 3.

34. Ron Broglio, *Beasts of Burden: Biopolitics, Labor, and Animal Life in British Romanticism* (Albany: State University of New York Press, 2017), 124.

35. Robinson, *Original Frankenstein*, 25, 29.

36. Robinson, 31.

37. Ashton Nichols, *Beyond Romantic Ecocriticism: Toward Urbanatural Roosting* (New York: Palgrave Macmillan, 2011), 38. See *The Frankenstein Notebooks*, ed. Charles E. Robinson, 2 vols. (New York: Garland, 1996), 1:85.

38. Nichols, *Beyond Romantic Ecocriticism*, 28.

39. Percy Shelley, Preface to *Frankenstein*, 3.

40. Erasmus Darwin, *The Temple of Nature*; or, The Origin of Society. A Poem (London: J. Johnson, 1803), 1.1.227.

41. Martin Tropp, *Mary Shelley's Monster* (Boston: Houghton Mifflin, 1976), 15; Anne Mellor, *Mary Shelley, Her Life, Her Fiction, Her Monsters* (New York: Routledge, 1988), 54.

CHAPTER 6

Playing Devil's Advocate

DEFENDING THE CRIMINAL JUSTICE SYSTEM IN *FRANKENSTEIN*

L. Adam Mekler

The title of this chapter reflects two of the primary problems that arise during consideration of Mary Shelley's first novel. The first is that discussion of the investigation and prosecution of the murders committed by the Creature in *Frankenstein* almost exclusively groups all of those crimes into one context, despite the fact that they occur in three different countries (Switzerland, Ireland, and France) and therefore are not subject to the same laws and investigational practices.[1] Together, these cases entail the integration of numerous factors—including nationality, religion, class, and gender—over and above the actual evidence involved. The complexity of this group of concerns leads to the second primary problem, which is the tendency to ignore the evidence and focus more exclusively on the other factors: Justine is convicted because she is a poor woman, for example, while Victor is acquitted because he is a rich white man, such investigation reveals.[2]

It is this tendency to see the justice system (or systems) as biased that engenders a desire to come to its "defense." Of course, it would be naïve, at best, to suggest that gender does not play a pivotal role in the novel or that the novel does not offer legitimate avenues for the critique of oppressive social systems. Indeed, much important discussion of the role of gender, especially, in the events of the novel in general and in the criminal trials in particular has been produced in recent years, filling a relative void in *Frankenstein* criticism that persisted up until the turn of the twenty-first century.[3] This study, however, explores the ways in which other interpretive possibilities may (co)exist. Indeed, although justice is not served equally well for Victor, for Justine, and, perhaps more importantly, for the Creature, the problems that arise for them in the novel are not solely the result of gender, class, or religious prejudice. A close consideration of these issues, of course, is necessary to an understanding of the presentation of justice in the novel. However, focusing exclusively on such factors runs the danger of ignoring important elements of the novel that contribute to Shelley's overall argument.

The first factor, therefore, that needs to be considered in this respect is what Switzerland, France, and Ireland meant to Great Britain during the period from 1816 to 1817, when Shelley was writing the novel, as well as in the late 1790s, during which time the novel is set.[4] France, of course, had been an adversary to England since the Norman Conquest, so the declaration of war between the countries in 1793 served to reignite hostilities that had been simmering for centuries, boiling over most notably in the Hundred Years' War (1337–1453), the Seven Years' War (1756–1763), and the American Revolution (1775–1783, with full French involvement officially beginning with the Treaty of Alliance in 1778). The commencement of the French Revolution in 1789 brought the rivalry between the two countries again to the fore, as prominent intellectuals and common folk alike debated the merits and dangers of the events across the Channel.

Although not as acutely influential as France at this time, Switzerland, and Geneva in particular, also evoked significant consideration by Romantic-era Britons. Shortly after the commencement of the French Revolution, Geneva had overthrown its ancien régime and established itself as a republic, though it was annexed by the French in early 1798 and did not reclaim its independence until Napoleon's defeat in 1813. It entered into the Swiss Confederation in 1815, only shortly before the Shelleys' arrival. Indeed, Geneva held considerable interest for both Percy and Mary Shelley during their 1816 visit. Percy, for instance, was particularly intent on seeing sites associated with Jean-Jacques Rousseau, whose works, including *The Social Contract* and *Emile*, both published in 1762, were influential to revolutionaries in the American colonies and France.[5] Mary also expresses her respect for Rousseau in her description of Plainpalais in *History of a Six Weeks' Tour* and in *Frankenstein*.[6]

Like France and Switzerland, Ireland also occupied a complex position in the minds of Britons during the time of the novel's setting and composition. Unlike the other two countries, however, Ireland was indeed part of the United Kingdom, following the Act of Union in 1800. Of course, the joining of the Kingdom of Ireland to that of Great Britain was an act more of political expediency during wartime; Ireland had been a point of invasion by France in 1798 and continued to be a feared target of the French. If anything, the Act of Union exacerbated the tensions between England and Ireland, primarily the result of religious differences, as the issue of Catholic Emancipation was particularly divisive. As William Gibson explains, William Pitt in fact "wanted to ease the Union's acceptance in Ireland by making concessions to the Catholics. This included granting Catholics some political rights."[7] When, in February 1801, George III refused to agree to such concessions, however, Pitt resigned. All of these political, military, and religious considerations find expression in Mary Shelley's novel.

Indeed, by the time of the novel's composition, and in the wake of the Napoleonic Wars, Great Britain had emerged as a victor in many ways. France had been returned to its 1790 boundaries; Britain's colonial presence had been solidified in such places as Malta, southern Africa, and Ceylon; and a substantial step toward the universal abolition of the slave trade had been achieved.[8] However, the glory of Britain's military successes was qualified by the political costs that the country was forced to face. Desperate to prevent the spread of revolutionary disease into its own land during the early 1790s, England clamped down on the civil liberties of its citizens to an alarming degree. Habeas corpus was suspended in 1790, in 1794, and again in 1817, being restored for good shortly after the publication of Shelley's novel. The Treason Trials of 1794 were another very clear example of the British government's extreme efforts to prevent insurrection, as twelve of the leading radicals of the time—including Thomas Hardy, John Horne Tooke, and John Thelwall—found themselves arrested, jailed, and tried on capital charges because of their political sentiments. Consequently, these legal developments provided substantial motivation for many of the writers of the late eighteenth century, in whose works David Punter identifies "a consistent discrediting of English legal mechanisms and institutions, and concomitant speculation about other systems, whether these are located abroad or in internal subcultures."[9]

One of the most prominent political writers in this regard, of course, was William Godwin, Mary Shelley's father and, significantly, the person to whom *Frankenstein* was dedicated when it was first published in 1818. In fact, his *Cursory Strictures on the Charge Delivered by Lord Chief Justice Eyre*, published in October 1794, was instrumental in helping turn public opinion in favor of the Treason Trial defendants, who eventually secured acquittal. In his *Things as They Are; or, The Adventures of Caleb Williams*, published earlier that year, the title character, a servant to the aristocratic Lord Falkland, becomes a victim of the British criminal justice system, which Godwin portrays as little more than the vehicle by which the rich maintain their political and economic power over the poor and laboring classes. As he explains, "Wealth and despotism easily know how to engage [the] laws as the coadjutors of their oppression."[10] Throughout the novel, Godwin indeed portrays a series of affluent characters exploiting the laws to confirm and exercise their own power, first in the description of Tyrrel's harassment of his ward and niece Emily Melville, later in his persecution of his tenant Mr. Hawkins, and finally in the extended narration of Mr. Falkland's oppression of the titular Caleb Williams.[11]

While Godwin's novel focuses on this conflict between the different classes, another novel written during the last decade of the eighteenth century explores the importance of gender to the foundations of the legal system. *Maria; or, The Wrongs of Woman* was left unfinished by Mary Wollstonecraft at her death in 1797, ten days after she had given birth to her daughter, Mary. Despite its incomplete status, however, William Godwin included it in his edition of the *Posthumous Works of the Author of "The Rights of Woman,"* and consequently it, along with Godwin's *Caleb Williams*, was well known to the young Mary Godwin, who was particularly moved by Wollstonecraft's portrayal of the main character's confinement in a mental hospital, the result of her depraved husband's efforts to secure the fortune Maria has attempted to keep from him.[12] Understandably, the laws of England, which essentially viewed a woman as the property of her husband, appear to have aided his efforts as they are portrayed at the end of the completed portion of the novel Wollstonecraft left behind.[13] Following the presentation of Maria's narrative before the court to defend her lover from the charges of seducing her, the judge summarily dismisses her arguments, making no mention of the several examples she has presented of her husband's misappropriation of her fortune, proclaiming only "the fallacy of letting women plead their feelings, as an excuse for the violation of the marriage-vow."[14] As would be seen later in Elizabeth's futile testimony during Justine's trial in *Frankenstein*, women's status in the courtroom, in both England and Switzerland, was indeed considered problematic, but Mary Shelley's portrayal of the role of the gender would be a bit more complex.

When Mary Shelley began *Frankenstein* in Geneva during the summer of 1816, she was understandably influenced both by the political developments of the previous twenty-five years and by her parents' literary responses to the society that produced those developments. Again, especially logical are the connections that can be made between the fate of women in the novel and the writings of Mary Wollstonecraft. More significant in this context, however, is the setting of each of the murders in a different country, which can be seen as an attempt to explore the alternate legal systems to which Punter refers. One such alternative system would have been found in Geneva, where the first murder trial occurs, resulting in the conviction and execution of Justine Moritz for the murder of Victor Frankenstein's youngest brother, William.

Interestingly, however, those critics who fully examine Justine's trial itself tend to refer only indirectly to the political context of the novel's late eighteenth-century setting. William Sayres, for example, declares, "we would recognize the jury trial in the enlightened republic of Geneva as approaching our own standards of burden of proof and presumptions of innocence," and consequently, he associates Mary Shelley's novel with William Godwin's *Caleb Williams*—thus conflating the Genevese and British criminal justice systems—by suggesting that in both works, "the charge of ingratitude subverts the laws completely, and sweeps aside the constraints of reason in legal process to convict both [Caleb and Justine] of crimes for which they were framed."[15] To be sure, the perception that Justine has been an ungrateful servant to the Frankenstein family certainly contributes to the community's ill will toward her. After Elizabeth testifies on Justine's behalf, as Victor recalls, "a murmur of approbation was heard; but it was excited by her generous interference, and not in favour of poor Justine, on whom the public indignation was turned with renewed violence, charging her with the blackest ingratitude" (I:VII.60).[16] That the victim's own adopted sister, emotionally asserting her belief in Justine's innocence, proves so unsuccessful, however, does not necessarily indicate that public opinion is a deciding factor in the trial's outcome.

In fact, Justine's conviction might confirm the importance of reason, rather than the subversion of reason, in this trial. Indeed, the evidence against Justine may ultimately be so compelling that no other verdict could logically be returned. As Victor's brother Ernest relates to him, Justine has taken ill after the discovery of William's body, "and, after several days, one of the servants, happening to examine the apparel she had worn on the night of the murder, had discovered in her pocket the picture of [Victor's] mother, which had been judged to be the temptation of the murderer. The servant instantly shewed it to one of the others, who, without saying a word to any of the family, went to a magistrate; and, upon their deposition, Justine was apprehended" (I:VI.56–57). While the servants' behavior may not appear to be the most compassionate response to the discovery of the locket—one wonders why they do not at least ask Justine about it—from a legal point of view, their actions are quite laudable.[17] Eliminating potential questions about the chain of custody that might compromise the integrity of the evidence, they take it directly to the authorities, who are judged best able to evaluate the implications of its discovery.

At the trial, the locket becomes the central piece of evidence against Justine. The other facts—including her inability to explain her whereabouts on the night of the murder and her confused response to the woman who questions her—do not carry nearly as much weight. Unfortunately, even Justine's own testimony confirms the locket's significance: "I know," she explains, "how heavily and fatally this one circumstance weighs against me, but I have no power of explaining it. . . . I believe that I have no enemy on earth, and none surely would have been so wicked as to destroy me wantonly. Did the murderer place it there? I know of no opportunity afforded him for so doing; or if I had, why should he have stolen the jewel, to part with it again so soon?" (I:VII.59–60). Justine's sentiments are understandable, certainly, but they do her no good here. Without being able to explain how she came into possession of the locket, she leaves herself with little hope for acquittal. Her final reliance on the testimony of character witnesses is also proven ineffective, as "fear, and hatred of the crime of which they supposed her guilty" prevent them, with the sole exception of Elizabeth, from coming forward (60). Of course, as William Veeder and Anne Frank Wake have both pointed out, even Elizabeth's testimony is itself compromised, as her speech reveals the imprint of her recognition of the differences—socioeconomic as well as familial—that exist between them.[18] Especially damning in this regard is Elizabeth's assertion, "For my own part, I do not hesitate to say, that, notwithstanding all the evidence produced against her, I believe and rely on her perfect innocence. She had no temptation for such an action: as to the bauble on which the chief proof rests, if she had earnestly desired it, I should have willingly given it to her; so much do I esteem and value her" (60). Reminding the court of the amount of evidence against Justine certainly does not bolster her defense. Further, as worded, Elizabeth's suggestion that she would have given the necklace to Justine reflects more on herself than on Justine: if, for example, she had stated that Justine *knew* that she could have had the locket if she had asked for it, her testimony may have been more helpful.

In the end, then, the judges are presented with little evidence to persuade them in Justine's favor and substantial, albeit circumstantial, evidence against her. As Beth Newman describes, "the evidence that links Justine to the murder is circumstantial, which means that it assumes a narrative form: a series of apparently related events is distributed into a pattern of cause and effect, and so into a single, coherent plot; this plot, being narratable, is plausible, and being plausible begins to seem true."[19] Therefore, despite our knowledge that Justine is in fact innocent, it is inappropriate to suggest that the system itself is unfair. Feminist critiques of the novel often validly offer Justine's case as evidence that Mary Shelley is protesting the failure of women to enjoy public power in a patriarchal society—see, for example, Anne K. Mellor's "Possessing Nature: The Female in *Frankenstein*"; and William Veeder suggests that "the machinery of the law" is against Justine, but Veeder must also admit that "Justine has been victimized by the very plot of the novel. How could she imagine that an eight-foot-tall, man-made monster had sneaked up and slipped the miniature into her pocket?"[20] Similarly, Newman explains that the Creature "is the one circumstance of which judge, jury, and Justine are alike ignorant, the missing cause that would make Justine's story hold together."[21] Justine has no way of knowing that the Creature exists, and neither do the judges. They can therefore judge her only on the evidence before them.

In this regard, the responses of both Victor and Elizabeth to Justine's conviction and execution, while providing some foundations for the indictment of the justice system in Geneva, appear a bit incongruent. When Victor first tells Elizabeth of the verdict, he proclaims that "all judges had rather that ten innocent should suffer, than that one guilty should escape" (I:VII.61). When the cousins visit Justine and hear that her confession has been coerced, Elizabeth responds more vehemently:

> Yet heaven bless thee, my dearest Justine, with resignation, and a confidence elevated beyond this world. Oh! how I hate its shews and mockeries! when one creature is murdered, another is immediately deprived of life in a slow torturing manner; then the executioners, their hands yet reeking with the blood of innocence, believe that they have done a great deed. They call this *retribution*. Hateful name! When that word is pronounced, I know greater and more horrid punishments are going to be inflicted than the gloomiest tyrant has ever invented to satiate his utmost revenge. (62)

After Justine's death, Elizabeth's emotions are a bit more subdued, yet her impression of society is only slighted softened:

> When I reflect, my dear cousin, . . . on the miserable death of Justine Moritz, I no longer see the world and its works as they before appeared to me. Before, I looked upon the accounts of vice and injustice that I read in books or heard from others as tales of ancient days or imaginary evils; at least they were remote and more familiar to reason than to the imagination; but now misery has come home, and men appear to me as monsters thirsting for each other's blood. (II:I.69)

Here, Victor and Elizabeth share a perspective that is both melodramatic and hyperbolic, to be sure, but also one that is not clearly supported by the other evidence Victor provides in his narrative. In each quote, the speaker depicts the justice system as so obsessed with revenge that it blindly pursues punishment, regardless of the facts of the case. Elizabeth's thoughts are especially harsh, describing the agents of the law as bloodthirsty tyrants, intentionally and gleefully inflicting pain on their helpless victims. Here, Wake suggests, "Elizabeth boldly articulates her frustration and anger toward the injustice of patriarchal constructs implied by the behaviors of the clergy, the jury, and . . . her fiance."[22] Wake also suggests that Mary Shelley "addresses, in abstract and subliminal language, unequal distributions of power, and the injustice of patriarchal systems, institutions, and customs."[23]

The problem here, unfortunately, is that, while the actions of Justine's confessor may justify, to some extent, Elizabeth's perspective especially, as well as Wake's interpretation of that perspective, Victor's description of the judicial process itself offers little such support. In fact, the agents of law in Victor's story are quite reasonable in their approach to Justine's prosecution. They allow Justine to speak in her own defense, for example, and indulge, albeit unsuccessfully, her request to call witnesses to testify on her own behalf. Even after Victor learns not only that Justine has been condemned but that she has also confessed, the officer who shares this news expresses no real joy. Victor relates, "That evidence . . . was hardly required in so glaring a case, but I am glad of it; and, indeed, none of our judges like to condemn a criminal upon circumstantial evidence, be it ever so decisive" (I:VII.61). These words hardly express a bloodthirst feeding on the suffering of the innocent.

If any group reveals a lust for vengeance, it is the people of Geneva themselves. As Elizabeth continues her discussion with Victor, she admits, "Yet I am certainly unjust. Everybody believed that poor girl to be guilty; and if she could have committed the crime for which she suffered, assuredly she would have been the most depraved of human creatures" (II:I.69). Although she falls just short of sharing such a mentality, Elizabeth does recognize some validity to the desire to exact revenge on those who commit especially egregious crimes. In this respect, then, the spectators at Justine's trial, even though they are not officials of the court, can be seen to exert some influence on the outcome of the trial, if not as much as the evidence itself, in their expression of indignation at Justine's perceived ingratitude.

In this regard, historical parallels can be found in both Paris and Geneva. During the summer of 1794, for example, an uprising occurred in Geneva that was at least partially, Fred Randel explains, instigated by France and that led to the formation of a Revolutionary Tribunal: "Under the influence of intimidation by the 'savage multitude,' and without credible judicial proceedings or evidence of violation of law, according to Invernois, the Tribunal executed eleven persons, including at least four magistrates, two of whom were ex-syndics or presidents of Geneva."[24] Echoing Burke's famous condemnation of the "swinish multitude" of the French Revolution and its early proponents in Britain, Invernois identifies in the populace of Geneva the same fervent desire for retribution that inspired the Women's March on Versailles in 1789 and the September Massacres of 1792, with comparably bloody consequences.

The combination of unusual circumstances and historical connections that inform the presentation of Justine's conviction will also be central to the next murder case, following the Creature's murder of Henry Clerval. This case develops after Victor has traveled to Scotland to fulfill his promise to his Creature to build him a companion. Ultimately deciding not to continue, he clandestinely sets out in a boat to dump the remains of his project, falls asleep, and ends up on the shore of Ireland, where he is apprehended for Clerval's murder. As in Justine's case, there is some compelling evidence against the accused: he is believed to be seen in a boat near the part of the shore where the body has been found; he is even more agitated than Justine is described as being; and he essentially confesses to the murder when he sees Clerval's body: "Have my murderous machinations deprived you also, my dearest Henry, of life?" he asks. "Two I have already destroyed; other victims await their destiny, but you, Clerval, my friend, my benefactor—" (III:IV.136).

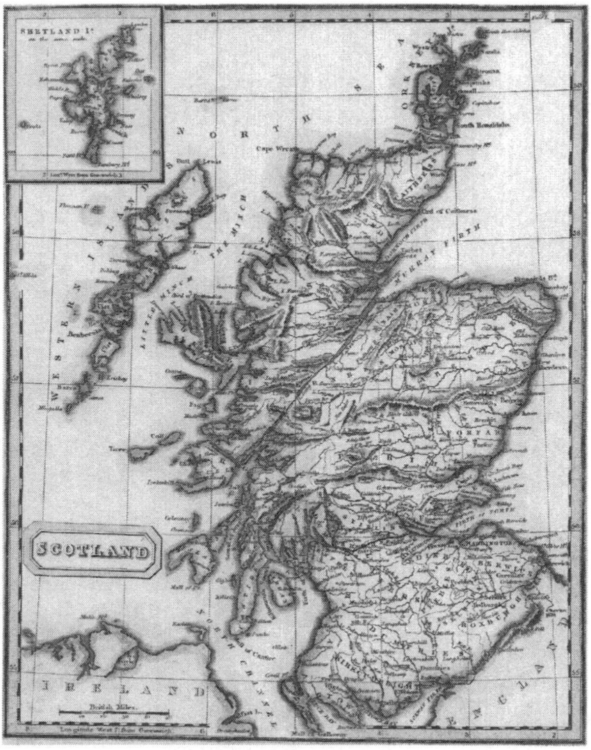

Figure 6.1. Map of Scotland with inset of Shetland Isles, ca. 1810, with Orkneys at top right, Ireland at bottom left. (Thomas Kelly, London)

Despite these developments, however, two essential evidentiary circumstances are again sufficient to determine the case's outcome. Unlike in Justine's case, there is no physical evidence that clearly links Victor to the crime. Further, unlike Justine, Victor is able to find people to testify on his behalf and thereby provide an alibi for him; subsequently, the grand jury rejects the bill against him when he is proven to have been on the Orkney Islands, over five hundred miles away, when Clerval's body is found (III:IV.140). Again, the plot of the novel itself arises as a determining force in the prosecution of Victor. If Veeder can ask how Justine could imagine that an eight-foot-tall monster has framed her, one must also be able to ask how it is at all possible that Victor could ever have ended up in Ireland from the Orkneys without seeing some other part of Scotland's coast (see figure 6.1).[25] Of more immediate concern, though, is the relative unlikelihood of Victor's indeed conversing with anyone on the evening of Clerval's murder. Victor has spent the afternoon preparing for the disposal of the remains of the female creature, and while he does mention having spent the afternoon sitting on the beach cleaning his "chemical apparatus" (III:III.132), he has earlier proclaimed that he "lived ungazed at and unmolested" (III:II.127). Regardless, on one level, the criminal justice system in Ireland works here in the same way as it does in Geneva. Given the evidence with which the magistrates are presented, they make the most logical determination. It is not immediately relevant that Victor is no more innocent in reality than Justine is or, apparently, that justice appears to be served better in the politically unstable Ireland than in Geneva, celebrated for its enlightenment philosophy and republican ideals.

One important difference, therefore, between these two cases lies in the actual investigation of the crimes. Specifically, while the Genevan magistrates do not compel Justine's character witnesses to testify on her behalf, Victor's alibi witnesses are apparently willing or at least more successfully encouraged to come forward on his behalf. In this situation, admittedly, class may indeed play a factor. Early in Mr. Kirwin's investigation, he tells Victor, "But you will, I hope, soon quit this melancholy abode; for, doubtless, evidence can easily be brought to free you from the criminal charge" (III:IV.138). Given the circumstantial evidence before him, it appears a bit unclear what Kirwin is basing this conclusion on, other than, perhaps, the dress and manner of the accused. He relates, "It was not until a day or two after your illness that I thought of examining your dress, that I might discover some trace by which I could send to your relations an account of your misfortune and illness" (139). As Victor's clothing would likely suggest a level of affluence greater than that of most of the townspeople, Kirwin may be predisposed to treat him with greater respect. Nevertheless, the more immediate explanation lies in the likelihood that Kirwin, and by extension the Irish system he represents, is more thorough in its investigational practices than are the Genevan magistrates in the earlier case.

In the case of the final murder of the novel, that of Elizabeth Lavenza Frankenstein, the criminal justice systems of both France and Switzerland are presented as even less effective. In this instance, the historical relationship between France and England in the 1790s offers some important context. As Randel suggests, the introduction of Evian allows Mary Shelley to establish a contrast between the independent republic of Geneva and the despotic kingdom of Sardinia, whose control over the town of Evian and the surrounding southern Lake Genevan shore had been restored by the Congress of Vienna, ending France's control of the region, which had lasted from 1792 until 1815. Randel explains, "For the Shelleys in 1816–1818 the Kingdom of Sardinia was a distillation of the most reactionary politics of the European Restoration."[26] Percy Shelley's description of the town in *History of a Six Weeks' Tour* makes this position clear: "The appearance of the inhabitants of Evian is more wretched, diseased and poor, than I ever recollect to have seen. The contrast indeed between the subjects of the King of Sardinia and the citizens of the independent republics of Switzerland, affords a powerful illustration of the blighting mischiefs of despotism, within the space of a few miles."[27] Thus, Randel is accurate in identifying the significance of Evian at the time of the novel's composition. Its status during the time period during which the novel is set, however, would necessarily be different, as at that time it was to some degree more an extension of the nascent French Republic and thus better associated with the extremes of the French Revolution.

In this context, Mary Shelley's setting of Elizabeth's murder here allows her to establish a connection to her earlier portrayal of injustice in France found in the stories of Safie, whose Turkish father is unfairly incarcerated by the French government, and of Felix's family, who are themselves unjustly imprisoned after he successfully helps the Turk to escape (II:V.91–93). As the Creature tells Victor,

> The father of Safie had been the cause of their ruin. He was a Turkish merchant, and had inhabited Paris for many years, when, for some reason which I could not learn, he became obnoxious to the government. He was seized and cast into prison the very day that Safie arrived from Constantinople to join him. He was tried, and condemned to death. The injustice of his sentence was very flagrant; all Paris was indignant; and it was judged that his religion and wealth, rather than the crime alleged against him, had been the cause of his condemnation. (91)

Unlike Justine's case, there is no discussion of the strength of the evidence against the Turk—or even, importantly, what the exact charges against him are—and it is the possession of wealth rather than its absence that is described as an influential factor. In any event, his case allows Mary Shelley to use the adversarial relationship between France and Great Britain as a backdrop to the novel's events. The outrage that Britons expressed in response to the reported excesses of the French Revolution, beginning of course with Edmund Burke's *Reflections on the Revolution in France*, led to a more positive comparative view of British justice, though this view was quickly changed, as we have seen, by the government's clamping down on civil liberties in the 1790s.

In this regard, then, although Elizabeth's murder in the Frankensteins' honeymoon suite in Evian does not explicitly evoke such considerations, the subtle associations continue to impact the novel's meaning. Here, Victor faces perhaps the strongest evidence of any of the accused in the novel: the servants hear Elizabeth's screams; when they go to the Frankenstein room to investigate, they find her dead on the bed with Victor passed out on the floor. After he awakens, he returns to the room and fires a gun, causing the people of the inn also to return to the murder scene and then to go outside to track the supposed killer. After several futile hours, Victor gives up, while, he admits, "most of my companions believ[ed] it to have been a form conjured by my fancy" (III:VI.151). Shortly thereafter, he returns to Geneva, where he loses his fragile hold on his sanity and is confined in a cell for several months. When, after recuperating, Victor finally goes to the local magistrate to tell his story, he is met with skepticism: "He had heard my story with that half kind of belief that is given to a tale of spirits and supernatural events; but when he was called upon to act officially in consequence, the whole tide of his incredulity returned" (153). The question then arises: If no one believes his story of the Creature, how else can they account for the murder of Elizabeth except by focusing on her husband, who could certainly be considered mentally unbalanced enough to have committed such a crime?[28] As with the resolution of the investigations into William's and Clerval's murders, Mary Shelley's explanation is not explicit, though it is important to recognize that Victor refuses to allow the legal system to complete its investigation before he embarks on his own quest for revenge.

The true accomplishment of justice, in Victor's eyes at least, must occur outside the scope of the legal system. Jonathan Grossman suggests in his discussion of the legal aspects of *Frankenstein* that the final two legal scenes "are less concerned with unveiling the injustice of the legal system than with revealing the inadequacy of the entire Western justice system, civil and criminal, to Frankenstein and the human creature's predicament."[29] In Grossman's view, which interestingly conflates the Irish and Genevan judicial systems, the real issue behind the different legal developments involves the private negotiations between Victor and the Creature regarding their proper relationship to each other, a procedure echoing the chancery suit that occupied much of the Shelleys' attention at the time of the novel's composition.[30] However true this may be, the fact remains that the interaction between Victor and his Creature is the closest the Creature comes to participating in any sort of legal procedure.

Indeed, from the beginning of the Creature's narrative, he reveals that his perspective on human legal systems has been mixed. When he first learns about the laws of society, as the result of observing Felix's instruction of Safie using Volney's *Ruins of Empires*, the Creature exclaims, "For a long time I could not conceive how one man could go forth to murder his fellow, or even why there were laws and governments; but when I heard details of vice and bloodshed, my wonder ceased, and I turned away with disgust and loathing" (II:V.89). Reinforcing his earlier claim to have been born with a soul that "glowed with love and humanity" (68), the Creature here demonstrates a naïve conception of the true nature of violence, as well as the need for humans to restrain their baser impulses by the imposition of laws. At the same time, the Creature repeatedly confirms his status as outside these laws. As Lisbeth Chapin also observes in her contribution to this volume, he learns about Volney, for example, only through eavesdropping on Safie's lessons. Later, when he finds a suitcase full of books (II:VII.95), his education regarding society's laws continues, but he gets no closer to acceptance by that society.

In this context, the Creature's repeated interactions with humans may be metaphorically identified as trials in that various individuals pass judgment on the Creature on the basis of their interpretation of the physical evidence literally before them—the Creature's actual appearance. In this way, the Creature's interview with the elder De Lacey proves pivotal in many ways, especially because this experience acts as the first of his trials in which he is actually able to speak on his own behalf and be heard.[31] On one level, De Lacey, as a blind man, symbolizes the ideal tribunal for the Creature. As he tells De Lacey, "I trust, that by your aid, I shall not be driven from the society and sympathy of your fellow-Creatures." De Lacey's response—"Heaven forbid! Even if you were really criminal" (II:VII.94)—offers the Creature hope that he can receive a fair hearing, but he finds out all too soon that human justice is in fact not really blind at all. When he hears Felix and the others approaching, the Creature pleads with De Lacey, "Do not desert me in the hour of trial!" but De Lacey can only stand by powerless and confused as Felix, like the other villagers who sit in immediate judgment of the Creature, imposes his own sentence by attacking the Creature in response to his hasty interpretation of the prone Creature's appearance before the elder De Lacey (94).

In the final analysis, this interaction represents the only trial in the novel—albeit a figurative rather than official one—that is truly unfair. Moreover, the importance of the fact that these are not actual legal trials should not be overlooked. Unlike in the cases of William, Clerval, and Elizabeth, here the defendant does not have the opportunity to provide a full defense before judgment is passed. Significantly, the Creature's male gender does not provide him any advantage at all, while his impoverished status also appears to be irrelevant. The discrepancies between this instance and the other, more official judicial proceedings therefore generally confirm the integrity of the larger systems, despite their apparent flaws.

Having established this distinction between the informal and formal trials in the novel, one must nevertheless acknowledge that a more beneficial outcome, at least temporarily, follows the Creature's first interaction with Victor after Victor's return from Ingolstadt. At first, Victor echoes the responses of the other humans by passing judgment on the Creature based on his literal appearance alone. As he sits in the mountains upon first returning to Geneva following William's murder, he captures a glimpse of the passing Creature:

> A flash of lightning illuminated the object, and discovered its shape plainly to me; its gigantic stature, and the deformity of its aspect, more hideous than belongs to humanity, instantly informed me that it was the wretch, the filthy dæmon, to whom I had given life. What did he there? Could he be (I shuddered at the conception) the murderer of my brother? No sooner did that idea cross my imagination, than I became convinced of its truth. . . . Nothing in human shape could have destroyed that fair child. *He* was the murderer! I could not doubt it. The mere presence of the idea was an irresistible proof of the fact. (I:VI.54–55)

As every other human has done before, Victor responds to the sight of his Creature by immediately attributing murderous inclinations to him. The fact that

Victor is accurate in his assessment despite the glaring logical fallacies actually serves, ironically perhaps, to establish a distinction between Victor and the Creature's interactions, especially their subsequent dialogue on Montanvert, and the Creature's experiences with other people.

Indeed, during that exchange, Victor becomes the first person to listen to the Creature's full defense of his actions, albeit reluctantly. The Creature tells him, "Listen to my tale: when you have heard that, abandon or commiserate me, as you shall judge that I deserve. But hear me. The guilty are allowed, by human laws, bloody as they may be, to speak in their own defence before they are condemned. Listen to me, Frankenstein. You accuse me of murder; and yet you would, with a satisfied conscience, destroy your own creature. Oh, praise the eternal justice of man!" (II:II.75). Granted, Victor's previous knowledge of the Creature gives him insight that the other people who have encountered him have lacked and thus makes him slightly more amenable, but the Creature nevertheless offers a logical plea for justice from his creator. Victor admits, as he follows the Creature to hear his tale, "I weighed the various arguments that he had used, and determined at least to listen to his tale. I was partly urged by curiosity, and compassion confirmed my resolution" (76). Just as the Creature has hoped, his acquisition of language has given him the ability to influence people's opinion of him,[32] and Victor's willingness to hear his full story opens the way for a fuller presentation of his testimony.

However, Victor's wavering responses after hearing the Creature's tale offers no clear indication that the interests of justice are being served. Repeatedly vacillating among anger, pity, doubt, and compassion, Victor approximates a judge or juror by weighing all the evidence as he deliberates before delivering a verdict:

> I paused some time to reflect on all he had related, and the various arguments which he had employed. I thought of the promise of virtues which he had displayed on the opening of his existence, and the subsequent blight of all kindly feeling by the loathing and scorn which his protectors had manifested towards him. His power and threats were not omitted in my calculations: a creature who could exist in the ice caves of the glaciers, and hide himself from pursuit among the ridges of inaccessible precipices, was a being possessing faculties it would be vain to cope with. After a long pause of reflection, I concluded, that the justice due both to him and my fellow-creatures demanded of me that I should comply with his request. (II:IX.110)[33]

Although the eventual outcome of this encounter appears to be a positive one for the Creature, the tremendous ambivalence that Victor expresses reveals that providing justice to him is not his sole, or perhaps even his primary, consideration. The desire to protect his "fellow-creatures" is certainly an understandably strong one, as is his recognition of the apparent futility of rejecting the Creature's plea.

Therefore, it is not at all surprising that Victor will eventually prove unable to carry out the agreement established under these circumstances.

Nevertheless, Victor's ultimate failure to carry out his promise, coupled with the similarly problematic conclusions to the other trials in the novel, confirms to some degree Bridget Marshall's claim about the novel: "*Frankenstein* . . . simultaneously places the reader in the position of juror and denies the reader sufficient evidence and testimony to make a sound decision in the case." Similarly, John Beer suggests that "the novel retains its potency by very reason of the fact that the questions it raises are unresolved."[34] The indeterminacy that Marshall and Beer perceive in the novel—the failure to develop a fully informed conclusion "in the case"—speaks to another avenue by which to explore Mary Shelley's concerns in writing it, one that actually moves, at least partially, beyond the consideration of the novel's historical and geographical contexts.

Indeed, some of the difficulties discussed earlier in trying to make sense of the novel's portrayal of justice may reflect a larger philosophical issue. In this context, the connections between Mary Shelley's novel and the life and works of Samuel Taylor Coleridge offer important implications. Most evident in this regard are Mary Shelley's references to Coleridge's *The Rime of the Ancient Mariner*. Other works, especially *Paradise Lost*, may receive more discussion in the novel,[35] but Coleridge's poem stands out more prominently given the anachronistic nature of its inclusion, having been published after the events of the novel are likely to have occurred.[36]

Two fairly recent studies of this connection have indeed argued that Coleridge's influence on the novel is more central than has traditionally been believed. In the essay "*Frankenstein* and *Christabel*: Intertextuality, Biography, and Gothic Ambiguity," Leslie Ann and Walter S. Minot argue that Coleridge is in fact a prototype for Victor Frankenstein, while Beth Lau argues in "Romantic Ambivalence in *Frankenstein* and *The Rime of the Ancient Mariner*" that, "far from being a target of disapproval, Coleridge was a profoundly sympathetic and congenial figure to Mary Shelley whose beliefs, themes, and literary techniques resonated with and helped shape her own."[37] One particularly significant common theme that Lau identifies is the ambivalent portrayal of the crime and punishment in the two works. Lau explains, "The ostensible moral of *The Ancient Mariner* and *Frankenstein*—that solitary self-assertion is a crime that should and will be punished—is . . . undermined by a number of elements that qualify the overreacher's guilt, diminish his responsibility for his crime, or otherwise confuse the work's point of view toward the central characters and their actions."[38] Here, Victor's crime in rejecting his Creature immediately upon his animation seems as inexplicable as the Mariner's killing of the albatross, with the eventual consequences of each appearing to be incommensurate with the crimes.

Beyond this shared theme, Mary Shelley makes the connection between these two texts more explicit through her use of allusions to Coleridge's poem at three key moments in the novel. Specifically, at the beginning of his narrative, Walton assures his sister, "I am going to unexplored regions, to 'the land of mist and snow;' but I shall kill no albatross, therefore do not be alarmed for my safety" (I:Letter II.14). Later, as Victor recalls wandering the streets after fleeing from his creation, he also refers to the poem:

> My heart palpitated in the sickness of fear; and I hurried on with irregular steps, not daring to look about me:
>
> Like one who, on a lonely road,
> Doth walk in fear and dread,
> And, having once turn'd round, walks on,
> And turns no more his head;
> Because he knows a frightful fiend
> Doth close behind him tread. (I:IV.41)

Finally, when Victor is contemplating his impending marriage to Elizabeth, he echoes Coleridge's juxtaposition of weddings and transgressions. Just as the Ancient Mariner's telling of his tale prevents the Wedding Guest from enjoying the wedding banquet, Victor's decision to participate in his experiments requires him to postpone his own wedding celebration:

> Alas! to me the idea of an immediate union with my cousin was one of horror and dismay. I was bound by a solemn promise, which I had not yet fulfilled, and dared not break; or, if I did, what manifold miseries might not impend over me and my devoted family! Could I enter into a festival with this deadly weight yet hanging round my neck, and bowing me to the ground. I must perform my engagement, and let the monster depart with his mate, before I allowed myself to enjoy the delight of an union from which I expected peace. (III:I.117)

In each of these passages, the use of Coleridge's poem helps to establish a sense of ambiguity. In the first quote, Walton's declaration seems a little illogical, as, despite his suggestion, the dangers faced by the Ancient Mariner and his fellow crew members by no means commence with the death of the albatross, despite its deadly aftermath. In the second passage, the source for which Mary Shelley makes sure to identify in a footnote,[39] Victor, like the Ancient Mariner, identifies himself not as a criminal but rather as a potential victim of a crime, despite the fact that Victor has abandoned an ostensibly helpless being for whom he should be taking responsibility. In the final passage, Victor's reference to the weight of the albatross around his neck is inserted into a scene that blurs the distinctions between his two obligations, to Elizabeth and to the Creature. Although the syntax indicates that it is his pledge to the Creature that weighs on him, the placement of the allusion to a "solemn promise" so close to the discussion of his other engagement clearly invites additional associations.

Given these parallels, the struggle to make sense of each work may also be examined along similar lines. Frances Ferguson asserts, "The difficulty of the poem is that the possibility of learning from the Mariner's experience depends upon sorting that experience into a more linear and complete pattern than the poem ever agrees to do."[40] In this reading, the attempts of critics to establish a coherent, unified message in the poem reveal the same frustrations and ambiguities that Marshall and Beer identify in *Frankenstein*. Coleridge addresses such concerns in his famous discussion of an exchange with Anna Laetitia Barbauld:

> Mrs. Barbauld once told me that she admired the "Ancient Mariner" very much, but that there were two faults in it,—it was improbable, and had no moral. As for the probability, I owned that that might admit some question; but as to the want of a moral, I told her that in my own judgment the poem had too much; and that the only, or chief fault, if I might say so, was the obtrusion of the moral sentiment so openly on the reader as a principle or cause of action in a work of such pure imagination. It ought to have had no more moral than the "Arabian Nights" tale of the merchant's sitting down to eat dates by the side of a well, and throwing the shells aside, and lo! a genie starts up, and says he *must* kill the aforesaid merchant, *because* one of the date shells had, it seems, put out the eye of the genie's son.[41]

While ostensibly agreeing with Barbauld that the poem's morality is suspect, Coleridge's admission has the larger effect of casting a skeptical view on the whole process of imposing morals at all. Ferguson argues, "As is common in Coleridge's work generally, intention and effect are absolutely discontinuous, and the moral is that morality appears to involve certainty only if you can already know the full outcome of every action before you commit it."[42] Of course, such logic goes, the Mariner would not kill the bird if he knows that two hundred men will die as a result, just as the merchant would not discard the date shells if he knows doing so will lead to the blinding of the genie's son or, more importantly, his own death.

Applying this lesson, then, to the presentation of justice in Mary Shelley's novel requires the exploration of one final parallel between Coleridge's poem, or more specifically his discussion of it found in *Table Talk*, and the novel, or more specifically Percy Shelley's Preface to it. There, Percy Shelley proclaims, "I am by no means indifferent to the manner in which whatever moral tendencies exist in the sentiments or characters it contains shall affect the reader; yet my chief concern in this respect has been limited to the avoiding the enervating effects of the novels of the present day, and to the exhibition of the amiableness of domestic affection, and the excellence of universal virtue."[43] While, of course, Percy Shelley would not have been able to read Coleridge's *Table Talk*, first published fourteen years after his death, he was able to read the lines that come near the conclusion of the *Ancient Mariner*:

He prayeth best who loveth best,
All things both great and small:
For the dear God, who loveth us,
He made and loveth all.

To be sure, this message, on its surface, is a positive one, but it is disconnected from what occurs in the poem. Percy Shelley's celebration of the domestic affections is similarly incongruent with the action of the novel. While in Coleridge's poem the failure to love one of God's creatures leads to the deaths of hundreds of people, in *Frankenstein* the failure fully to maintain one's appreciation of one's family leads to their destruction. In both texts, the lessons are almost comically inappropriate.[44]

Ultimately, the representatives of the different criminal justice systems in the novel are confronted by similar difficulties. Unable, for example, to take into account the existence of the Creature in considering either Justine's or Victor's guilt in the first two murders, they must base their conclusions on the evidence before them. While the Creature's presence is at least discussed during the investigation into Elizabeth's murder, other important factors, including Victor's own mental instability, must also be considered. Further, the failure of these different systems to bring the actual murderer, the Creature, to justice, does not necessarily reflect the existence of intentional bias on their part but rather indicates the difficulty, in the midst of so many different factors, seen and unseen, of establishing a perfect mechanism by which to assess and administer justice at all.

NOTES

1. See, for example, Jonathan Grossman, *The Art of Alibi: English Law Courts and the Novel* (Baltimore: Johns Hopkins University Press, 2002), 75; Colleen Fenno, "Monstrous Crimes and Offender Accountability: Shelley's Representation of Criminal Justice from *Frankenstein* to *Falkner*," in *Critical Insights: Mary Shelley*, ed. Virginia Brackett (Ipswich, MA: Salem, 2016), 150–151; and Bridget Marshall, *The Transatlantic Gothic Novel and the Law: 1790–1860* (Farnham, UK: Ashgate, 2011), 69.

2. See, for example, Anne K. Mellor, "Possessing Nature: The Female in *Frankenstein*," in *Romanticism and Feminism*, ed. Anne K. Mellor (Bloomington: Indiana University Press, 1988), 221; and Fenno, "Monstrous Crimes and Offender Accountability," 151.

3. In addition to the Grossman, Fenno, and Marshall texts previously cited, especially insightful considerations also include Ann Frank Wake, "Justine's Trial Revisited: A Space for Women's Subculture in Mary Shelley's *Frankenstein*," *European Romantic Review* 12, no. 4 (2001): 493–516; and Patrick Vincent, "'This Wretched Mockery of Justice': Mary Shelley's *Frankenstein* and Geneva," *European Romantic Review* 18, no. 5 (2007): 645–661. Lisbeth Chapin and Lisa Crafton also consider the importance of gender in their essays for this collection.

4. For an important discussion of the issue of geography in the novel, see Fred Randel, "The Political Geography of Horror in Mary Shelley's *Frankenstein*," *ELH* 70, no. 2 (2003): 465–491. Although Randel focuses on the murders themselves more than the investigations and trials associated with them, his study offers important consideration of the political

context for the events of the novel and is indeed helpful in its discussion of the trial of Justine, as is discussed shortly.

5. Leo Damrosch explains, "It is always said that the American Founders were chiefly influenced by Locke and Montesquieu, and so they were. But some were influenced by Rousseau as well, though his radical reputation made it unwise to say so openly. Jefferson's immortal line, 'We hold these truths to be self-evident, that all men are created equal, that they are endowed by their creator with certain inalienable rights,' comes directly out of *The Social Contract*." Damrosch, "Friends of Rousseau," *Humanities* 33, no. 4 (2012): 12.

6. See, for example, Percy Shelley's letter to Peacock (dated 12 July 1816), which served as the basis for part of his contribution to *History of a Six Weeks' Tour*. In the letter, Percy presents a running account of his travels with Byron at the end of June, culminating with a description of their visit to Edward Gibbons's house: "My companion gathered some acacia leaves to preserve in remembrance of him. I refrained from doing so, fearing to outrage the greater and more sacred name of Rousseau; the contemplation of whose imperishable creations had left no vacancy in my heart for mortal things." In *Letters of Percy Bysshe Shelley*, ed. Frederick L. Jones, vol. 1 (Oxford: Oxford University Press, 1964), 487–488. Similarly, Mary Shelley explains, in her description of the Plainpalais in the *History*, "Here a small obelisk is erected to the glory of Rousseau, and here (such is the mutability of human life) the magistrates, the successors of those who exiled him from his native country, were shot by the populace during that revolution which his writings mainly contributed to mature, and which, notwithstanding the temporary bloodshed and injustice with which it was polluted, has produced enduring benefits to mankind." *History of a Six Weeks' Tour*, ed. Jeanne Moskal, vol. 8 of *The Novels and Selected Works of Mary Shelley* (London: Pickering & Chatto, 1996), 46.

7. William Gibson, *A Brief History of Britain: 1660–1851* (London: Constable and Robinson, 2010), 205; Susan Kroeg, "From 'This Shore, Renowned for Its Hospitality' to 'The Detested Shore of Ireland' in *Frankenstein*," *English Language Notes* 42, no. 3 (2005): 22.

8. See R. K. Webb, *Modern England*, 2nd ed. (New York: Harper, 1980), 155; and Gibson, *Brief History of Great Britain*, 210.

9. David Punter, "Fictional Representation of the Law in the Eighteenth Century," *Eighteenth-Century Studies* 16, no. 1 (1982): 47.

10. William Godwin, *Caleb Williams*, ed. Gary Handwerk and A. A. Markley (Peterborough, ON: Broadview, 2000), 137.

11. For the clear influence of Godwin's work, and this novel in particular, on Mary Shelley's composition of *Frankenstein* and other novels, see Pamela Clemit, *The Godwinian Novel: The Rational Fictions of Godwin, Brockden Brown, Mary Shelley* (New York: Oxford University Press, 1993), and William Brewer, *The Mental Anatomies of William Godwin and Mary Shelley* (Madison, NJ: Fairleigh Dickson University Press, 2001).

12. John Williams, *Mary Shelley: A Literary Life* (New York: St. Martin's, 2000), 35.

13. As William Blackstone asserts in his *Commentaries on the Laws of England*, "By marriage, the husband and wife are one person in law: that is, the very being or legal existence of the woman is suspended during the marriage, or at least is incorporated and consolidated into that of the husband: under whose wing, protection, and cover, she performs every thing; and is therefore called in our law-French a *feme-covert*, . . . is said to be *covert-baron*, or under the protection and influence of her husband, her *baron*, or lord; and her condition during her marriage is called her *coverture*." See *Blackstone's Commentaries Abridged*, ed. William Sprague (Chicago: Callaghan, 1915), 82. It would not be until the Married Women's Property Acts of 1870 and 1882 that women would obtain the right to their own possessions.

14. Mary Wollstonecraft, *The Wrongs of Woman; or, Maria*, ed. Janet Todd and Marilyn Butler, vol. 1 of *The Works of Mary Wollstonecraft* (New York: New York University Press, 1989), 181.

15. William Sayres, "Compounding the Crime: Ingratitude and the Murder Conviction of Justine Moritz in *Frankenstein*," *English Language Notes* 31, no. 4 (1994): 49, 50. While essentially valid, however, Sayres's assertions are not completely accurate. In *Caleb Williams*, for example, Caleb is actually never convicted of any crime, though he is imprisoned for long periods of time, is perpetually dogged by accusations of criminal activity, and has numerous encounters with the machinery of the law.

16. Quotations follow the publication of 1818, as presented in *Frankenstein; or, The Modern Prometheus*, ed. Nora Crook (London: Pickering & Chatto, 1996).

17. Fenno offers an interesting counter perspective in her essay, arguing that the circumstantial evidence against Justine is "manipulated" by a system that disadvantages "unprotected defendants." Fenno, "Monstrous Crimes and Offender Accountability," 148–149. However, the only explicit and intentional manipulation that Fenno discusses is that by the Creature—certainly not himself an agent of any social or political institution—in placing the locket in Justine's possession.

18. See William Veeder, *Mary Shelley and "Frankenstein": The Fate of Androgyny* (Chicago: University of Chicago Press, 1986), 174–177; and Wake, "Justine's Trial Revisited," 497–498.

19. Beth Newman, "Narratives of Seduction and the Seductions of Narrative: The Frame Structure of *Frankenstein*," *ELH* 53, no.1 (1986): 148.

20. Veeder, *Mary Shelley and "Frankenstein,"* 183.

21. Newman, "Narratives of Seduction," 150.

22. Wake, "Justine's Trial Revisited," 501.

23. Wake, 502.

24. Randel, "Political Geography," 470, quoting Francois d'Invernois, *A Short Account of the Late Revolution in Geneva* (London: Spilsbury, 1795), 30.

25. Leslie Klinger also points out the difficulty Victor would have had in traveling so far in such a short period of time, suggesting instead the Inner Hebrides as a more logical point of origin. See *The New Annotated Frankenstein*, ed. Leslie S. Klinger (New York: Liveright, 2017), 212.

26. Randel, "Political Geography," 485–486.

27. Quoted in Randel, 485.

28. Of some potential importance in this context is the apparent similarity between Victor's account of his encounter with the Creature in the bridal chamber and Percy Shelley's controversial account of an apparent assassination attempt in his lodgings at Plas Tan-Yr-Allt in Wales on February 26, 1813. Not only can parallels be drawn between the particulars of the two incidents—including the firing of guns in rooms without witnesses, followed by the incredulous responses of several of the people who respond to the commotion—but there are also interesting connections along legal lines. Richard Holmes, for example, identifies the creation of "Associations for the Prosecution of Felons" in response to increasing tension in the area at the time, which included occurrences of "Feloniously Breaking and Entering any Dwelling House, in the night time." Thus, the Creature's crimes of trespassing and murder apparently have some immediate historical parallels. See Holmes, *Shelley: The Pursuit* (New York: Dutton, 1975), 187. Holmes discusses the Plas Tan-Yr-Allt episode fully on 187–188 and 190–197. For more explicit connections to *Frankenstein*, see Holmes, 332.

29. Grossman, *Art of Alibi*, 75.

30. Grossman, 65–66.

31. Michael Scrivener makes his identification of fictional trials a bit broader in his discussion of the portrayals of trials in literature of the era, declaring the entire novel itself to be "in some sense an extended trial, with Walton justifying the deaths of the sailors because of his ambition, with Frankenstein defending his murderous dissecting of nature, and with the creature pleading his innocence as a murderer." Scrivener, "Trials in Romantic-Era Writing: Modernity, Guilt, and the Scene of Justice," *Wordsworth Circle* 35, no. 3 (2004): 131.

32. I refer, of course, to the Creature's subsequent reference to his conception of language as a "godlike science" in response to his observation of the De Lacey family's conversations (II:IV.83) and his hopes eventually to win them over with his "wisdom" (II:VII.98). For further consideration of the persuasive intent of the novel's first-person narratives, see Brian Bates's chapter in this collection.

33. In Stuart Curran's annotated edition of *Frankenstein* on the Romantic Circles website, he offers an interesting footnote to a reference Victor makes as he considers the justice of the Creature's argument. Curran suggests, "Perhaps the Creature comprehends that is he talking to a person from a long line of civic magistrates, or perhaps his plea has simply been predicated on their shared Enlightenment discourse. Whatever the case, in his long narration he has consistently been touching on the nerve from which he at last gets a response. For the first time in his dealings with the Creature, from the original thought of his creation down to this moment of moral decision, Victor acknowledges an abstract concept of the justice owed him." *Frankenstein*, ed. Stuart Curran, Romantic Circles, accessed July 7, 2021,https://romantic-circles.org /editions/frankenstein/V@notes/justice2.

34. Marshall, *Transatlantic Gothic Novel*, 70; John Beer, "Mary Shelley, *Frankenstein*," in *A Companion to Romanticism*, ed. Duncan Wu (Oxford, UK: Blackwell, 1998), 235.

35. These other texts include the works of Agrippa and Paracelsus read by Victor and the other texts read by the Creature, such as Plutarch's *Lives*, Goethe's *The Sorrows of Young Werther*, and Volney's *Ruins of Empires*.

36. For consideration of the time frame for the novel, see especially Charles Robinson's introduction to *The Frankenstein Notebooks: A Facsimile Edition of Mary Shelley's Novel, 1816–17 (with Alterations in the Hand of Percy Bysshe Shelley)*, 2 vols. (New York: Garland, 1996); and Essaka Joseph, "'Marking the Dates with Accuracy': The Time Problem in Mary Shelley's *Frankenstein*," *Gothic Studies* 3, no. 3 (2001): 279–308.

37. Leslie Ann and Walter S. Minot, "*Frankenstein* and *Christabel*: Intertextuality, Biography, and Gothic Ambiguity," *European Romantic Review* 15, no. 1 (2004): 24; Beth Lau, "Romantic Ambivalence in *Frankenstein* and *The Rime of the Ancient Mariner*," in *Fellow Romantics: Male and Female British Writers, 1790–1835*, ed. Beth Lau (Burlington, VT: Ashgate, 2009), 74.

38. Lau, "Romantic Ambivalence," 86.

39. Of the nearly dozen or so direct quotes from other literary works in the novel itself (excluding, therefore, the epigraph from Milton's *Paradise Lost* on the novel's title page), *The Ancient Mariner* is one of only three that I believe Shelley explicitly identifies (the others being Leigh Hunt's "Story of Rimini" and Wordsworth's "Tintern Abbey").

40. Frances Ferguson, "Coleridge and Deluded Reader: 'The Rime of the Ancient Mariner,'" *Georgia Review* 31, no. 3 (1977): 620.

41. Samuel Taylor Coleridge, *Table Talk*, in *The Best of Coleridge*, ed. Earl Leslie Griggs (New York: Ronald, 1945), 535–536.

42. Ferguson, "Coleridge and Deluded Reader," 624.

43. Preface to *Frankenstein*, 8.

44. In contrast, Percy Shelley's subsequent attempt to explain the novel, his 1817 review published posthumously in 1832, offers a more logical lesson, identifying the Creature's crimes as "the children, as it were, of Necessity and Human Nature. In this the direct moral of the book consists. . . . Treat a person ill, and he will become wicked." See Percy Shelley, "On *Frankenstein*," in *Frankenstein: A Norton Critical Edition*, ed. J. Paul Hunter (New York: Norton, 2012), 214. For Samuel Taylor Coleridge's poem, see "The Rime of the Ancient Mariner" in *Coleridge's Poetry and Prose*, ed. Nicholas Halmi, Paul Magnuson, and Raimondo Modiano (New York: Norton, 2004), 98.

CHAPTER 7

TEACHING *FRANKENSTEIN* AS PASTICHE, PARODY, AND ADAPTATION IN THE GENERAL EDUCATION CLASSROOM

Brian Bates

Charles Robinson's groundbreaking scholarship in *The Frankenstein Notebooks* reminds us that even though the novel presents Frankenstein's creation of the Creature as a solitary act, collaborative authorship was (and remains) central to the novel's composition and fiction-making constructs: "What Percy did to (or for) the novel is 'figured' by both Walton and Victor Frankenstein as editors of each other's narratives. Indeed, the more we look at the novel, the more we are led to the conclusion that *Frankenstein* is a series of texts in search of an editor, one who will ultimately give form and shape to the novel."[1] To my mind, Robinson's insight about the novel's searching openness to editorial reshaping describes the vital spark that has given it so many afterlives over the past two hundred years.[2] In one of Robinson's last published essays, the introduction to *Frankenstein: Annotated for Scientists, Engineers and Creators of All Kinds* (2017), he connects the novel's lability to our twenty-first-century focus on a science, technology, engineering, arts, and mathematics (STEAM) education philosophy, and the contextual essays that follow this new edition of the 1818 *Frankenstein* maintain that its continuing relevance in the arts and sciences depends on our own acts (as teachers and students) of invention.[3] Building on this spirit of shared invention, in this essay I reflect on a classroom experiment in the Spring of 2018 during which students collaborated with the novel to create a pastiche or parody representing what they found to be most alive in *Frankenstein*.[4]

I carried this experiment out in a general education Masterworks of British Literature course at California Polytechnic State University, San Luis Obispo.

Cal Poly is highly regarded for its engineering, architecture, and agriculture programs, and it prides itself on a "learn by doing" philosophy, which encourages students to focus on innovation while applying knowledge to real-world experiences. With that STEAM philosophy in mind, I asked students to create a pastiche or parody adaptation of *Frankenstein* that fills in one of the novel's textual gaps, elisions, ellipses, or unexplained secrets and then write a short preface describing the purpose of their creation and the particular intratextual conventions that they reanimated. Through their additive imitations of the novel's epistolary form, narrative point of view, and character motives, they played with and on our shared twenty-first-century assumptions about the generative power of coauthorship and adaptation as viable artistic technologies of innovative replication.

We identified and unpacked those assumptions on the second day of class while discussing Guillermo Del Toro's 2018 *Shape of Water* British Academy of Film and Television Arts (BAFTA) acceptance speech in relation to a five-minute, YouTube mashup of *Frankenstein* movie scenes—both of which highlight Mary Shelley's novel as an enduring and vital modern cultural myth that continues to inspire adaptive, multimedia creations involving pastiche and parody.[5] After positing working definitions of "adaptation," "pastiche," and "parody," we read the 1818 version of *Frankenstein* with an eye toward how the 1818 Preface and 1831 Introduction foreground these very concepts as signal parts of its multilayered authorship. Following that close textual engagement, we read the first stage adaptation of *Frankenstein*—Richard Brinsley Peake's successful melodrama *Presumption; or, The Fate of Frankenstein* (1823)—and began theorizing about how and why genre adaptations and textual transformations take shape in particular cultural contexts. That inquiry, which focused on melodrama, led us to several twentieth- and twenty-first-century multimedia adaptation pastiches and parodies, including the films *Frankenstein* (1931), *The Bride of Frankenstein* (1935), *Young Frankenstein* (1974), and *The Rocky Horror Picture Show* (1975) and Victor Lavalle's graphic novel series *Destroyer* (2017).

Students were intrigued by this approach to *Frankenstein*, but I also quickly realized that many of them had checked their "learn by doing" spirit of open inquiry and innovation at the door because the class was titled "Masterworks."[6] What they brought into the classroom instead were assumptions about what "great books" or "classic works" of literature are and how they should be read. As we discussed their assumptions, they cobbled together a definition of a literary masterwork as an old book, written by an ingenious single author, with a serious heroic or tragic action that teaches valuable, universal lessons. Faced with such masterworks in a classroom, they saw themselves as largely passive readers tasked with identifying and absorbing those lessons. Working against these notions of a masterwork, we approached *Frankenstein* as an open text that invites readers to adapt its very makeup. This method of reading to re-create helped to dispel their preconceptions about literary works as autonomous authorial creations and gave them permission to see themselves as active readers and inventive coauthors adding to *Frankenstein*'s legacy.[7] Moreover, it encouraged them to explore and play on their preconceptions about high aesthetic masterworks versus popular culture productions—all of which the 1818 Preface and 1831 Introduction to *Frankenstein* cannily educe.

We read the 1818 Preface and 1831 Introduction as paratexts, inviting readers to consider *Frankenstein*'s novelty as a function of its collaborative authorial creation, adaptation aesthetics, and modes of pastiche and parody.[8] From this perspective, Percy Shelley's 1818 Preface styles the anonymously published novel less as the work of a single author (the Preface uses "I") than as a tale of combined authorial influence. From the very first sentence, which introduces "Dr. Darwin, and some of the physiological writers of Germany," Percy describes *Frankenstein* as a pastiche and genre adaptation of other authors' ideas and conventions. He aligns the novel with and separates it from popular culture genres—"mere tale[s]" of "spectres or enchantment"—and grants "prose fiction a license, or rather a rule," to "innovate" by adapting from "the *Iliad*, the tragic poetry of Greece,—Shakespeare, in the *Tempest* and *Midsummer Night's Dream*—and most especially Milton, in *Paradise Lost*," the most "exquisite combinations of human feeling," which "have resulted in the highest specimens of poetry" (4).

The final two paragraphs of the Preface further obfuscate notions of single authorship and original genius. They directly refer "to the author" of *Frankenstein* and describe "the summer of 1816 spent in Geneva" with literary friends, but they also describe the novel's origins as a joint writing venture responding to "some German stories of ghosts, which happened to fall into [the authors'] hands": "These tales excited in us a playful desire of imitation" (3–4). Percy characterizes this story-writing compact as an amusing imitation game, and the word "playful" suggests that the friends' "agree[ment] to write each a story" involved comedic or satiric parody as well as pastiche. The final paragraph of the Preface plays on the ephemerality of their story creations—two of these friends lost "all memory of their ghostly visions"—and leaves readers with *Frankenstein* as the only surviving representation of their collective endeavor: "The following tale is the only one which has been completed" (5). According] to the Preface, *Frankenstein* came into being as an adaptation—a prose-poetry hybrid, folktale pastiche, ghost-story parody, and literary record of communal authorship.

In the 1831 Introduction, Mary Shelley plays a double game that asserts her authorship of *Frankenstein* while also defining that authorship according to a set of adaptation practices (in dialogue with the 1818 Preface and the novel's reception history) vacillating between pastiche and parody. Just over halfway through her Introduction, she winks at the very idea of singular authorial creation: "Invention, it must be humbly admitted, does not consist in creating out of a void, but out of chaos; the materials must, in the first place, be afforded: it can give form to dark shapeless substances, but cannot bring into being the substance itself. . . . Invention consists in the capacity of seizing on the capabilities of a subject, and in the power of moulding and fashioning ideas suggested to it" (189). Through her playful allusion to God's creation acts and Chaos in *Paradise Lost*, Mary presents imaginative authors not as original geniuses but as talented artisans and astute arrangers with the "capacity" to animate, by remaking, already existing "ideas" and "materials." Her "Sanchean" *Don Quixote* quip at the beginning of this paragraph, "Everything must have a beginning," and ensuing reference to the "story of Columbus and his egg" mock and further demystify original authorial creation.

These parodic twists redound two paragraphs onward when Mary recounts how she conceived of the idea for the novel in a dream. She describes her dream as a vision in which Victor created the Creature, escaped into a nightmarish sleep, and then woke to find his creation materialized at his bedside. Likewise, she frames her initial authorial act as a turn from a disturbing late-night conversation about reanimation into nightmarish sleep followed by a shocking eureka awakening that gives formal shape to her "hideous phantom." As a few of my students noted during class discussion, Mary's parallel between her own creation scene and Victor's seems a little too on the nose, particularly since she published her account thirteen years after the 1818 novel. They wondered which idea came first—Victor's creation scene or Mary's—and if Mary's 1831 recounting might be read as a tongue-in-cheek parody of authorial creation, signaled by her characterization of Victor's act as "mock[ing] the stupendous mechanism of the Creator of the world" (190).

With this spirit of open inquiry into slippery authorship, modes of creation, and textual doubling, we turned to the 1818 novel to discover and analyze the intratextual Gothic workings of its narrative points of view (including tone, diction, and syntax), plot organization (including dialogue and the pace of the narrative), and the motives of different characters. Through our close textual investigations, students determined the narrative rules that structure the novel's form and began considering whether they felt drawn to create a serious pastiche that imitates and extends part of that narrative form or a parody that pokes fun at particular aspects of the novel's first-person, epistolary frame narratives. To help catalyze that process of synthesizing and translating these structural elements into their own creations, much of our discussion focused on how Walton's, Victor's, and the Creature's first-person narratives each have a distinct persuasive purpose: the Creature is intent on persuading Victor to make him a female partner; Victor is intent on persuading Walton to kill the Creature if Victor dies first; and Walton is intent on persuading his sister, Margaret, of the high-minded nobility and societal benefits of his North Pole expedition. During our close analysis of each narrator's persuasive tone, diction, and syntax, students identified and evaluated the Gothic characteristics and power dynamics at work on the particular auditor/reader of each tale. They also examined how each narrator's manipulative, rhetorical stance indirectly reveals the overreaching, overdetermined, exaggerated, and/or fabricated parts of his narrative. Their responses to these power dynamics were mixed, ranging from terror at the devastating, single-minded power of these male narrators to seeing one or more of them as laughably ridiculous man-children.

Victor, the college student, garnered much of our attention. In the span of two years, he transforms from an uninformed and reluctant first-year student at the University of Ingolstadt into a self-driven, misanthropic autodidact whose technical innovations (and secret creation project) far outstrip the work of his teachers. While Victor's most conspicuous secreting of his technological innovations occurs in his underdescribed creation of the Creature, his enigmatic handiwork also shadows the novel's entire fiction-making technology and connects its patchwork construction with both of his Creatures: the seemingly complete male Creature and the never completed, mutilated female Creature. As if the novel's unreliable, triple-layered, nesting-doll-like, first-person narratives were not trying enough for students, they learned in the last section of the novel that Victor has ghostwritten parts of Robert Walton's journal account. Walton remarks, "Frankenstein discovered that I made notes concerning his history: he asked to see them, and then himself corrected and augmented them in many places; but principally in giving the life and spirit to the conversations he held with his enemy. 'Since you have preserved my narration," said he, 'I would not that a mutilated one should go down to posterity'" (III:VII.166). Victor's underexplained corrections and augmentations of Walton's notes highlight what we considered as the animating dynamic of the novel: its additive technology, which invites readers to replicate, correct, and augment its textual body. Whether in the guise of Victor's unseen corrections and additions or Margaret Walton Saville's unseen letter responses (among others), the novel encourages readers to imaginatively role-play and add to its incomplete form.[9]

When we turned next to Peake's melodrama *Presumption; or, The Fate of Frankenstein* (1823), I was pleasantly surprised to discover that most students were amused and/or intrigued by the novel's first stage adaptation. *Presumption*'s compression and rewriting of the novel's plot, tweaking of the novel's characters and their relationships, retrofitting of scenes and redeployment of the novel's lines, and introduction of Fritz as Frankenstein's comic assistant encouraged them to imagine how and why they might add to *Frankenstein* with their own textual reworkings. Our reading of Peake's *Presumption* alongside early reviews of the 1818 novel also helped students grasp what several reviewers lamented or berated: the novel's uneven style and mixed aesthetics, ambiguous purpose, and questionable genre. The ultraconservative *British Critic*'s April 1818 review offers

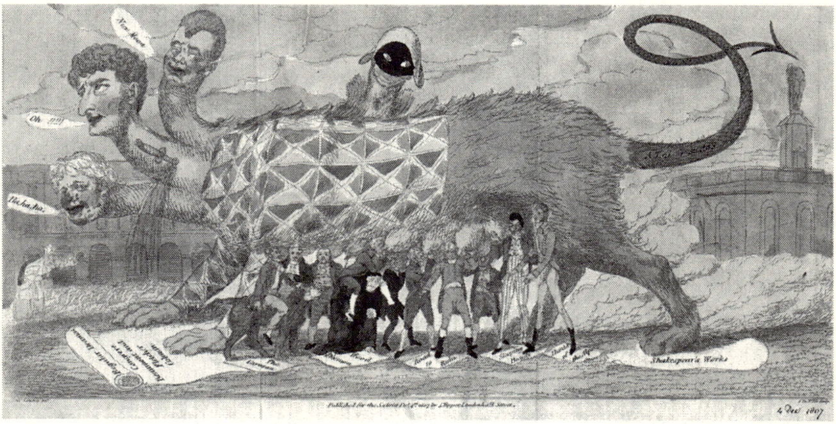

Figure 7.1. This print by Samuel De Wilde, "The Monster Melo-drama," from *The Satirist*, December 4, 1807, illustrates the concept of the monstrous melodrama.

the most extreme version of this critique: "We need scarcely say, that these volumes have neither principle, object, nor moral; the horror which abounds in them is too grotesque and *bizarre* ever to approach the sublime, and when we did not hurry over the pages in disgust, we sometimes paused to laugh outright; and yet we suspect, that the diseased and wandering imagination, which has stepped out of all legitimate bounds, to frame these disjointed combinations and unnatural adventures, might be disciplined into something better."[10] This reviewer describes *Frankenstein* as an illegitimate form, which does not know its own purpose and undoes its sublime reaching with grotesque language and exaggerated scenes that lead readers to veer wildly between "disgust" and "laughter."[11] Looking back at the novel through the lens of *Presumption*'s melodrama, my students began to see that this early review characterizes the novel itself as a monstrous melodrama (and an amateur one at best) from which the reviewer hopes the author "might be disciplined into something better"—a more purposeful mode of writing and/or a higher genre (figure 7.1).

Instead of a complete genre transformation, *Presumption* struck us as an adaptation bringing forth the novel's overt and latent melodramatic characteristics.[12] The play's superadditions of music and gesture also aided students in seeing how the machinery of an adaptation can add layers of meaning that enhance and critique the adapted text. Fritz embodied those additions with his triple roles as Frankenstein's comic foil, the play's rustic clown, and the audience's sometime surrogate. Fritz's comic duet with his wife, Ninon (French for "neither nor") in the final act also prompted us to theorize about melodrama's amorphous power to please, captured by the wordplay of "mellow" with the "melos" in melodrama. In their duet, Ninon accuses Fritz of being drunk, and he replies that he is "only mellow": "We'll all be so, for that is fun and life."[13] Occurring just before the Creature's kidnapping of William and the play's speeding toward a sensational finish, this tongue-in-cheek song celebrates gothic melo- ("mellow") drama's capacity to offer audiences an intoxicating, honeyed mix of low, bodily humor amid sweet (because fleeting) terror and overwrought pathos. Such melodrama's paramount stage trick involves working up the audience's emotions while also managing to leave them with Fritz's pleasant buzz and "mellow" feeling: "We'll all be so, for that is life and fun." In *Presumption*, that melodramatic "life" and "fun" appear most powerfully through the Creature's gestural responses to music and other characters.[14] In his first music-filled appearance, the Creature emerges through smoke and flame, breaks through a balustrade, "jumps on the table beneath, and from thence leaps upon the stage, [and] stands in attitude before *Frankenstein* who had started up in terror."[15] The Creature's dramatic pose not only displays his body (with blue-gray skin and Roman toga) but also presents him as a figure for the affective power of melodrama and its gloriously monstrous adaptations. From scene to scene thereafter, his mute expressions and pantomimic gestures imitate and conjure the highest and lowest of human passions. Whether the Creature is "snatch[ing] at the empty air" to grasp the fading music of Felix's flute, "cast[ing] himself at [Victor's] feet, imploring protection," or parodying Victor's "footsteps, making an ineffectual attempt occasionally to gain his loaded pistol," his movements are intertwined with the play's sudden, emotionally manipulative, musical turns.[16] *Presumption*'s unabashed focus on stimulating and guiding audience responses helped students to reflect on what their own purposes might be in adding to *Frankenstein*'s melodramatic adaptation legacy, which we explored next in the twentieth and twenty-first century.[17]

While a few of these students' adaptation pastiches and parodies were loose, patchwork constructions with vaguely drawn purposes and underdeveloped executions, the majority of their genre reframings and uses of pastiche and/or parody were sophisticated, incisive, and insightful. Several of their projects involved topics that align with this collection of essays—from Wolfson's, McCutcheon's, and Crafton's dynamic essays about science, technology, and media to Watters's, Chapin's, and Mekler's compelling essays about ecologies of food, climate, and social justice. About half the class wrote either a pastiche or parody imitating the novel's epistolary forms and narrative styles. The other half created multimedia adaptations. In the former half, many students wrote serious pastiches to fill in a perceived gap within the narrative (the creation scene most often) or to provide a brief prequel or sequel. Many of these pastiches represented the novel's underheard, muted, or overwritten female characters to give them a fuller voice, personal agency, and character complexity. Several other students wrote parodies imitating the novel's epistolary forms and narrative styles to poke fun at or strenuously ridicule Victor's toxic masculinity, overwrought emotionality, and tyrannical behaviors. A few students parodied particular scenes in the novel between male characters to highlight what they saw as their repressed or latent homoerotic relationships. The students who created multimedia adaptations incorporated a variety of twenty-first-century epistolary and social media modes, ranging from Twitter feeds and blogs to a version of PostSecrets.[18] Two students created abbreviated graphic novel pastiches, one student wrote a riotous Broadway musical parody, and another developed a single-player video game.

Charles Robinson probably would not approve of these additive, *Frankenstein*-inspired "hideous progeny." His meticulous and capacious editorial work might even suggest that he would dismiss such adaptive pastiches and parodies as trifling that only further muddies the waters of debate over Mary Shelley's authorship of *Frankenstein*. As Albert Rivero points out in "Whose Work Is It Anyway; or, How We Learned to Stop Worrying about the Author and Love the Text," Robinson privileges an author-centered (Mary Shelley) editorial method and is alarmed by the idea of a "hypertextual free-for-all" in which future readers can "assemble their own *Frankensteins*."[19] I want to suggest that both can happen—that as teachers we can convey to our students a robust fidelity to authorial *Frankenstein* texts and open up spaces for students to experiment with *Frankenstein* through collaborative reading and writing acts. Whether Robinson intended it or not, his scholarship on *Frankenstein* teaches us that if there are "masterworks," they are never finished and only continue to exist through our shared acts of imagining them anew. As I look back with gratitude on my conversations with Robinson over twenty years ago, I also realize that having the opportunity to pursue such open and attentive collaborative acts, particularly in the general education classroom, can be vital sparks leading us to discover who we are and how to make a place for ourselves in the world.[20]

NOTES

1. Charles Robinson, "Texts in Search of an Editor," in *Textual Studies and the Common Reader: Essays on Editing Novels and Novelists*, ed. Alexander Pettit (Athens: University of Georgia Press, 2000), 92.

2. On that history, see Chris Baldick, *In Frankenstein's Shadow* (Oxford, UK: Clarendon, 1987); Fred Botting, *Making Monstrous* (Manchester, UK: Manchester University Press, 1991); and, most recently, Dennis R. Cutchins and Dennis R. Perry, eds., *Adapting "Frankenstein": The Monster's Eternal Lives in Popular Culture* (Manchester, UK: Manchester University Press, 2018).

3. David Guston, Ed Finn, and Jason Scott Robert, the editors of *Frankenstein: Annotated for Scientists, Engineers, and Creators of All Kinds* (Cambridge, MA: MIT Press, 2017), include seven short essays that intertwine cross-disciplinary ways of thinking about *Frankenstein* with several of its most pressing contemporary contexts.

4. In contrast to Frederic Jameson's notion in "Postmodernism" (*New Left Review* 1, no. 146) of postmodern pastiche as dehistoricized and flattening, I follow Linda Hutcheon's contention in *A Theory of Adaptation* (New York: Routledge, 2006) that pastiche, parody, and adaptation can shine new light on history, literature, and our modes of critical evalua-

tion. Simon Dentith's *Parody* (London: Routledge, 2000) and Julie Sanders's *Adaptation and Appropriation* (London: Routledge, 2005)—both in the New Critical Idiom series—also provide useful historical surveys, definitions, and theories of parody, pastiche, and adaptation.

5. For Del Toro's February 18, 2018, acceptance speech, see BBC, "Guillermo Del Toro Wins Best Director BAFTA—The British Academy Film Awards: 2018—BBC One," YouTube, February 17, 2018, https://www.youtube.com/watch?v=UGHim_jC_jU; and for the YouTube mashup, see glasseyepix, "Fessenden Frankenstein Mashup," YouTube, March 1, 2012, https://www.youtube.com/watch?v=UA6zQbRlHoQ.

6. Because this is a general education class, I do not have license to change the title in the course catalogue.

7. Much of my thinking about active reading and writing draws on Roland Barthes's notion of writerly and readerly texts in *S/Z*, trans. Richard Miller (New York: Farrar, Straus and Giroux, 1974), and Umberto Eco's notion of open and closed texts in *The Role of the Reader* (Indianapolis: Indiana University Press, 1979).

8. For the 1818 Preface, 1831 Introduction, 1818 *Frankenstein; or, The Modern Prometheus*, Peake's 1823 *Presumption*, and early reviews of the novel, I quote from our course text, *Frankenstein: A Longman Cultural Edition*, 2nd ed., ed. Susan Wolfson (New York: Longman, 2007). This chapter follows the Longman edition; quotations from all these materials are from this edition; quotations from the preface, introduction, and novel are cited parenthetically in the text.

9. For more on these concepts, and a particular focus on Shelley Jackson's *Patchwork Girl* (Watertown, MA: Eastgate Systems, 1995) and Dave Morris's iPad *Frankenstein* app (2012), see Tully Barnett and Ben Kooyman, "Assembling the Body/Text: *Frankenstein* in New Media," in Cutchins and Perry, *Adapting "Frankenstein,"* 295–315.

10. See also *Frankenstein*, 376, 377, 382, 384, 390; the *Quarterly Review*'s January 1818 characterization of the novel as "a tissue of horrible and disgusting absurdity" marked by the reader's "struggle between laughter and loathing" (John Wilson Croker, *Quarterly Review* 18 [January (delayed until June) 1818]): 379–385); the Scots *Edinburgh Magazine*'s assessment of the novel's propensity for "caricature and exaggeration" with an "execution [that] is imperfect, and bearing the marks of an unpractised hand" (*Edinburgh Magazine and Literary Miscellany; A New Series of "The Scots Magazine"* 2 [March 1818]: 249–253); and the *Literary Panorama*'s evaluation that "the work seems to have been written in great haste, and on a very crude and ill-digested plan; and the detail is, in consequence, frequently filled with the most gross and obvious inconsistencies" (*Literary Panorama and National Registry* n.s. 8 [June 1, 1818]: 411–414).

11. I draw on Jane Moody's definitions of "illegitimate" in *Illegitimate Theatre in London, 1770–1840* (Cambridge: Cambridge University Press, 2000).

12. On the significance of melodrama as an abiding contemporary genre and pervasive mode of writing that emerged in the aftermath of the French Revolution, see Peter Brooks, *The Melodramatic Imagination* (New Haven, CT: Yale University Press, 1976). On melodrama's mixed forms, music, gesture, and capacity to transport audiences by manipulating the pace of their experiences, see Jeffrey Cox, "On the Death of Tragedy; or, The Birth of Melodrama," in *The Performing Century*, ed. Tracy C. Davis and Peter Holland (New York: Palgrave Macmillan, 2007), 161–181.

13. *Frankenstein*, 362.

14. For relevant scholarship on Peake's *Presumption*, see Emma Raub's "*Frankenstein* and the Mute Figure of Melodrama," *Modern Drama* 55, no. 4 (Winter 2012): 437–458; and Glen Jellenik's "A Frankensteinian Model for Adaptation Studies, or 'It Lives!': Adaptive Symbiosis and Peake's *Presumption, or the Fate of Frankenstein*," in Cutchins and Perry, *Adapting "Frankenstein,"* 45–61.

15. *Frankenstein*, 340–341.

16. *Frankenstein*, 346, 356, 366.

17. We gathered much of our energy for this brief (and necessarily incomplete) survey from *The Rocky Horror Picture Show*'s (1975) camp aesthetic.

18. See https://postsecret.com/.

19. Albert Rivero, "Response: Whose Work Is It Anyway? Or, How We Learned to Stop Worrying About the Author and Love the Text," in Pettit, 188. His essay responds to and concludes the collection of essays.

20. When I first met Charles Robinson in the spring of 1997, I was a junior at the University of Delaware looking for a professor to direct the honor's thesis that I wanted to write about John Keats. I had neither taken a class with him nor read any of his scholarship. To my distinct surprise, within minutes of our first meeting in Memorial Hall, he not only agreed to oversee my thesis but declared that I should write about Keats and Lord Byron and then whisked me through an open door to tour a Byron memorabilia room that he was putting together. Although I remember seeing a lock of Byron's hair, I was struck most by Robinson's seemingly indefatigable energy and capacity to draw me into a world of collaborative thinking. Over the next year, we met twice a month in his office to discuss the direction and progress of my thesis. Only recently have I begun to realize the extent to which these conversations influenced my teaching philosophy and practice.

Postscript

REMEMBRANCES OF CHARLES E. ROBINSON

Robin Hammerman

Charlie Robinson was an influential builder of community and a genuine friend to more colleagues than one could possibly name here. He guided his life's work with family in mind: his beloved wife, Nanette, was a regular fixture at the many conferences he regularly attended. His children, John and Clare, were the focal points of his joy, and he relished his role as grandparent. His dedication to the profession followed suit, to the extent that a global family of Romanticists blossomed under his care.

There is much to be gained in the remembrances contributed to this volume by Charlie's students and colleagues.[1] They collectively speak of a luminary figure whose steady hand and rigorous professionalism in his scholarship, teaching, and service to the academy were tempered with gracious humanity. This postscript reminds us that the work populating our orbit in conferences and classrooms, as well as on our bookshelves and databases, has extraordinary value beyond that of exchanging ideas. Charlie's unique legacy has already been recorded for posterity in several print and digital venues, including the memorials in 2017 by Susan J. Wolfson for the *Keats-Shelley Journal* and Jeffery Vail for the *Byron Journal*.[2] Those memorials affirm what Charlie would likely want us to communicate in a volume dedicated to his memory: that our business as humanists is about people.

The importance of Charlie's scholarship speaks for itself among those who study and teach Romantic-period texts. He graduated from Mount Saint Mary's College in Maryland and earned his PhD in 1967 from Temple University. He developed his dissertation, titled "The Frustrated Dialectic of Byron and Shelley: Their Reciprocal Influences," under the magnificent direction of David V. Erdman. Charlie joined the English Department at the University of Delaware in 1965, and this remained his academic home for the duration of his career. Lord Byron and Percy Shelley remained major figures of research for

Charlie's first published book in 1976, *Byron and Shelley: The Snake and Eagle Wreathed in Flight*, and he simultaneously produced *Mary Shelley: Collected Tales and Stories*.[3] Charlie continued his research on Byron and Percy Shelley while he devoted his attention to Mary Shelley, who hovered outside the canonical "big six" Romantic-period writers as a celebrity in her own right.[4] He edited two editions of collected essays on Byron: *Lord Byron and His Contemporaries: Essays from the Sixth International Byron Seminar* in 1982 and, in 1988 with Bernard Beatty and Tony Howe, *Liberty and Poetic License: New Essays on Byron*.[5] All this time, Charlie developed substantial work on the literature of Mary Shelley, subsequently publishing *The Mary Shelley Reader* in 1990 with Betty T. Bennett.[6] Two years later, he published (also with Betty Bennett) the facsimile edition titled *Mary Shelley's Mythological Dramas: "Proserpine" and "Midas."*[7]

Charlie's most widely known and celebrated publication is probably *The Frankenstein Notebooks*.[8] This two-volume folio edition features photographs of the manuscripts of *Frankenstein* with Charlie's transcriptions of the Shelleys' hands adjacent to the 1818 print pages. *The Frankenstein Notebooks* is an ambitious work of extraordinary editorial detail, from Charlie's transcriptions to his chronology and notes. Charlie traced his remarkable processes of discovery and evaluation in the notes, which alone resemble a substantial frame tale of enduring quality for those who study and teach *Frankenstein*. A follow-up edition, *The Original Frankenstein*, exemplifies his detail-oriented work ethic.[9]

Several works in progress beyond *Frankenstein* remained on Charlie's mind during the last months of his life, including an edition of William Hazlitt's correspondence and a volume on Percy Shelley's publisher, Charles Ollier. Charlie recovered Ollier's letters, and they held his interest for several decades. He left the files on Hazlitt and Ollier to trustworthy scholars, Hazlitt's to Duncan Wu and Ollier's to Michael Edson, so that they might bring his work to fruition. Still, Charlie never abandoned his best-known scholarly progeny, his work on *Frankenstein*.

Charlie's posthumously published introduction to the MIT *Frankenstein*, edited by David Guston, Ed Finn, and Jason Robert, ushered Mary Shelley's novel into the twenty-first century for "scientists, engineers, and creators of all kinds." The editors affirm that Charlie assumed more responsibility for this project than what it required from him. He put his whole heart into the development of this edition, and Charlie's efforts therein show us his effervescence as a scholar-teacher. Not only was he comfortable in this role, it seemed native to his core. Therefore, it makes sense that Charlie's last published work would reach what was, for him, a new student base. David Guston offers the following tribute on behalf of the editorial cohort behind this important new edition of the novel:

> I had the privilege of working with him up to the very end. His introductory essay to the edition of Mary Shelley's *Frankenstein* that I edited with ASU colleagues Ed Finn and Jason Robert was his last completed work before his death. In fact, I labored under the burden of knowing—fearing—that it might be. Charlie had confided in me about his condition, as our close working relationship over his essay required it. Such little bonds of secrecy help deepen relationships, moving colleagues to friends rapidly if, ultimately, painfully.
>
> Charlie prepared me for this shift by being such a generous and game collaborator. My fellow editors and I were on the receiving end of his experience with the genre and the text, being schooled gently about the origins of the use of the term "political science," the "nature of 'daemons,'" and so forth. But we did find and correct at least a couple of remaining errors in the corrected 1818 text that Charlie had so laboriously compiled, leaving what we published seemingly the best version available. And he acceded to substituting our "creature" for his "monster" to fit our editorial perspective and give our readers a bit more agency in assessing Victor's new, hybrid being. Our last exchanges, by email and by phone, involved the close details of his essay, his garden, and the impact of failing health on his ability to work both. I am pleased to say now, but to my consternation then, that the garden sometimes took precedent over the essay. I of course remain saddened that his death took precedence over both.[10]

Charlie's humanizing influence provided the first bits of nourishment for thousands of students, far more than just those who took his classes at the University of Delaware and the several fortunate ones who flourished under his direction for theses and doctoral dissertations. Contributions to the profession at large by his mentees seem practically endless. These accomplishments, obviously foremost to the credit of individuals, collectively stand as testimony of Charlie's dedication to their success. A sample of remembrances by students at his home institution, one for each decade of his career, provide a glimpse into Charlie's identity as a teacher.

Marsha Manns, cofounder with Leslie Marchand of the Byron Society of America, remembers Charlie's classroom in 1969. To Manns, every class was "an adventure, crammed full of passion and excitement." Students began to arrive in advance of the 8:30 A.M. start time "already talking eagerly about the readings for the day." It did not take long for Charlie to start coming early, too. He was, Manns recalls, eager to join in the preclass conversation. Moreover, "he lingered long after class to talk with entire groups of students—so long, in fact, that the incoming class had to push past the crowded doorway in order to make their way in."[11]

Lisbeth Chapin, whose essay appears in this volume, remembers Charlie as being "nothing short of a luminary" when she began her studies with him in

1976. "He venerated the work of Mary Shelley because he knew it was deserved, and I am a *Frankenstein* scholar to this day because of his devoted attention to that novel," Chapin's most memorable days in subsequent courses she took with him were about that fateful summer in Geneva, 1816. "In Charlie's classroom, the lightning crackled, the rain pattered down, and that galaxy of literary personalities clustered together on those dreary evenings sparkled in the candlelight. We weren't just in a literature classroom; we had entered a new universe of literature, a place of mountains, lakes, laboratories, and books, of ice and storms and waking dreams." Charlie's mentoring of Chapin was, to her, invaluable. He had, she remarks, "the greatest impact on all that I teach, research, read, and write." When Chapin was a graduate student, she retained the privilege of his mentorship as a reader of her dissertation on Percy Shelley when her own English department was two thousand miles away. Chapin remembers, "He was not easy to please; his scrutiny of syntax, word choice, and thesis was relentless, but because of that devotion I am a contributing member of the scholarly community today."[12]

Jeffery Vail recalls his first meeting with Charlie in 1987 when he was "a jejune (one of Charlie's favorite words) eighteen-year-old introvert about to graduate from Washington College": "I had applied to the University of Delaware's graduate program in English, and had driven from Chestertown, Maryland to Newark, Delaware for my interview." He remembers Charlie at that first meeting speaking loudly into the telephone, all the while in constant motion around his office. That energy was alive in his office, where Vail observed "glorious disorder everywhere."[13] Charlie became Vail's dissertation advisor, and their connection remained close for the rest of Charlie's life.

L. Adam Mekler, a contributor to this volume, recalls Charlie's "seemingly effortless way of guiding [students'] understanding of the material" in a senior seminar on Byron and the Shelleys during the summer of 1990: "He was so deliberate and matter-of-fact in his critique of our work, for example, that I couldn't fully appreciate his method—frankly, I found myself anxious to impress him in a way I wasn't with my other professors." They stayed in close touch throughout the years, and when Mekler was a graduate student elsewhere, Charlie continued to mentor him.

> We would occasionally meet in his office, which was overflowing with materials from the Byron Society archives while his tables were piled high with manuscript pages for what would eventually be *The Frankenstein Notebooks*. When I began teaching at Morgan State University after receiving my doctorate, it seemed like kismet that the resident Romanticist at the time had been close friends with Charlie since their time together in graduate school at Temple. When I assumed her position following her retirement, Charlie was of course one of the first people I told, and his congratulations upon my receiv-

ing tenure were particularly encouraging. In short, it is not coincidental that in many ways I have modeled myself as a professor after Charlie's example. Even so many years later, I continue to think of him almost every time I sign off on an email to a student in the same way he used to, simply providing my initials in lower case letters. In Charlie's case, though, that signature belied an importance that could never be diminished.[14]

Halina Adams, who worked on her dissertation with Charlie during his last years at the University of Delaware, echoes what others recognize of Charlie's teaching throughout his career: "he started the process of stitching together my academic identity." Adams describes her first meeting with Charlie on a "typical summer's afternoon" at the university: "Papers, books, and journals covered every surface, with a styrofoam cup appearing occasionally with Lilliputian mold islands floating on the dark coffee's surface. Behind the desk (and a couple of very unstable-looking towers of papers) sat a man in a blue vest, plaid button-down shirt, and khakis. This, I would come to learn, was his uniform. Today the vest was chalk-free, but when semester started in six weeks, one could chart how his day was going based on how many white smudges appeared on Charlie's signature blue sweater vest." Adams remembered moments from this meeting: "when I was transcribing a letter in Edinburgh ('you have to be careful, it's our job') or giving a conference paper in New York ('your prose should sing') or submitting my first article ('aim high, what have you got to lose?')." His voice remains with her: "as I pace about my classroom drawing strange diagrams on the board ('felix culpa!') and insisting on the real, living, necessary need for Romantic poetry ('If Winter comes, can Spring be far behind?')." To Adams, "Charlie's insistence on the importance of everything, every little detail was his way of making sure his knowledge lived on. His stories, his experience reading applications, his encyclopedic familiarity with jobs lists—it all needed to keep going. It was vitally important."[15]

Charlie went supremely above and beyond the typical call of duty with his students, and he did the same with his service to literary societies and professional organizations. Charlie's personable leadership skills were native traits. To him, it was a chief priority to be accessible in person, by telephone, and by written correspondence. His ebullient character is well documented in his emails. There, he was known to frequently use all-caps for emphasis, a hallmark that Adams calls "Robinsonian typographical flair." He devoted boundless energy to the Byron Society of America for several years as its executive director, and he did the same for other organizations including the Keats-Shelley Association of America, the Modern Language Association of America, and the Wordsworth-Coleridge Association. Through these venues, he actively fostered a sense of community among scholars, teachers, and enthusiasts. To him, this was one way of keeping the literature he loved (and sometimes loved to loathe) alive and thriving in the world.

Charlie nurtured scholars around the globe. For some, one meeting was all they had, and yet it was sufficient to sustain them. Anna Mercer, author of *The Collaborative Literary Relationship of Percy Bysshe Shelley and Mary Wollstonecraft Shelley* (2020), numbers among them. "Although I only met Charlie Robinson once, he has had a profound impact on my life." Mercer is "heavily indebted" to Charlie's work and to the advice he gave her when she came to Delaware from the United Kingdom to develop research for her book: "When I came to the USA to study Shelley manuscripts in 2015, Charlie met me in Delaware and the guidance he gave me on my research was invaluable; he was such a truly incredible scholar. He was so welcoming and dynamic. He was a real inspiration to me, and to many others. I have no doubt that Charlie was an excellent Ph.D. supervisor and I feel so grateful to have experienced some of that from across the Atlantic and then in our one meeting."[16]

Geoffrey C. Bond (OBE, DL, FSA), former chairman of the Byron Society in London, worked with Charlie on what he modestly calls "a number of matters" concerning the International Association of Byron Societies. He notes that "Charlie did much to extend and promote Byron Societies around the world." Bond, who also remembers Charlie as having a great sense of humor, offers the following anecdote:

> I remember him staying with us at Burgage Manor on one occasion when I had explained to him very carefully, knowing as a lawyer there is a propensity to litigate in the US, that he should use the slippers provided when crossing the wide wooden slippery hall floor of the Manor. The following morning, I was coming through the hall as Charlie came down the main staircase, no slippers, as his stockinged feet hit the hall floor he slipped and fell on his rear! I looked at Charlie and he looked at me, he smiled and said "mea culpa Geoffrey—no way I will litigate!"

Not only did the two work on matters concerning the International Association of Byron Societies, but Charlie also encouraged Bond to author his own work: "On one occasion when visiting Delaware, I had the pleasure of visiting the unusual community setting in which Charlie and Nanette lived. It was at this time I discussed my wish to write a book about Byron and his love of animals but was somewhat nervous of doing so. Without prompting Charlie told me to get on with it and not to worry about academia! I did subsequently write the book." Bond tenderly recalls that Charlie was "known throughout the Byron world as a scholar, friend and raconteur": "I can think of no better memorial to a person that they are remembered with the greatest affection in the memory of people. Charlie was a gentleman scholar who will be missed."[17]

Andrew Stauffer, president of the Byron Society of America, remembers Charlie this way: "He was a model for me personally and for so many others. Charlie studied, shared, and gave of himself; and he left us far too soon. But we have his brilliant scholarship, his field-changing editorial work, and his innumerable influences still visible in his students, his colleagues, and his friends." Stauffer observes that Charlie's degree of engagement or leadership remains unmatched: "[It] manifested itself in those familiar all-capitals emails that he would send to me, the board, and virtually every Byron enthusiast on his distribution list. He was a one-man hub of Romanticism, and I still miss those urgent, informative messages that would knit us all together." Stauffer recalls the International Association of Byron Societies' annual conference in August 2001, when Charlie was the amiable host. That year's meeting took place in the United States, and Charlie orchestrated it as a traveling show between Delaware, New York City, and Boston. Stauffer remembers Charlie at that time: "[He] command[ed] a busload full of international Byron scholars across a three-city tour, bringing us down from Boston to Delaware via Manhattan in the midsummer heat. I never saw him tired or anything less than jaunty and in control, as he marched us through our events with a precision signaled by frequent genial harangues on the details of the next episode. While the rest of us were wilting, Charlie was brimming with appetite for what lay ahead." The aftermath of that conference was when 9/11 occurred, and Stauffer remembers, "it was to Charlie that so many of us turned to convey news from the attendees, to share condolences, and to start to rebuild."[18]

Marilyn Gaull, who joined Charlie in the great beyond not long after she shared a remembrance of Charlie for this volume, was in New York City at that time. "After 9/11, when I was truly missing, Charlie recalled that I had a brother and using vast phone resources found him in Florida thinking he would know where I was. My brother was not even looking for me, so Charlie went on the hunt. That is another story. Charlie found me, and I have never been lost since, except when he passed."[19]

Hermione deAlmeida, a professor emerita of English at the University of Tulsa, observes, "To us ordinary mortals, fellow Romanticists, Charlie was the quintessential scholar's friend." In deAlmeida's estimation, "[Charlie was the] go-to person for answers and help on scholarly questions, however arcane, and he would invariably respond with the meticulous cheer that we all associate with his ongoing commitment to The Byron Society." Their occasional conversations in the afternoon were, to her, "a tonic." She remembers their last conversation in June 2016, when they spoke for a short time: "Charlie did not break character as the cheerful, concerned scholar and mentor—and he signed off, seemingly as always, with an 'OK, time to go, Hermione. You know how it is.' At that time, I did not know what he meant. We still have the legacy of Charlie's vitality—and his belief in a 'pay forward' community of scholars. His work remains our friend."[20]

Charlie extended a full heart and soul to his friendships with colleagues, a fact that Doucet Devin Fischer knows well. Her connection with Charlie spans several decades after they first met "long ago at the old Pforzheimer Library at

Madison Avenue and Forty-second Street." Fischer recalls that her "most vivid, and in a way heartbreaking" memory of Charlie concerns his generosity to the great Percy Shelley scholar Donald H. Reiman, who was for several decades at the editorial helm of *Shelley and His Circle* (its present incarnation is the Carl H. Pforzheimer Collection at the New York Public Library). Reiman, who also taught courses at the University of Delaware, died three years after Charlie to the day. Fischer suggests that Charlie "would have been the first to admit that Don Reiman was helpful to him," especially in the early years of his career. She continues,

> I suspect that only a few know that Don was repaid a hundredfold, as his mind was slowly but inexorably ravaged and his wife's health declined. It was Charlie who helped Don manage the gift of his valuable library to the University of Delaware; Charlie who enabled Don to sort through his papers; Charlie who took Don to lunch and drove him to doctors; Charlie who fielded frightened calls in the middle of the night; Charlie who visited Don during his frequent hospitalizations and served as his medical historian; Charlie who proved to be the catalyst for a welcome family rapprochement; and, finally, it was Charlie, when he was already fatally ill, who managed to find the Reimans a safe harbor in a Wilmington facility. If, as the old song goes, all you really need is heart, Charlie had heart in abundance.[21]

Those who worked side by side with Charlie at the University of Delaware know intimately the extent to which he was a center of intelligence, competence, and care. Donald Mell, the now-deceased professor emeritus of English at the University of Delaware and former director of the University of Delaware Press, remembered Charlie as "a great friend and colleague" from the time Mell arrived on the UD campus in the fall of 1968. He continued,

> I worked with him as secretary of the graduate committee of the English department for many years, and was interim director before his appointment, which turned out to be a brilliant decision on the part of Zack Bowen, the current chair of the department at the time. Charlie's contributions as board member and chair were stellar. He took on the job of vetting manuscripts that involved textual editing and bibliography accuracy and shaped them for publication as no other faculty member could possibly do with such skill—and all the while churning out articles on Byron and Shelley, writing a critical study covering both authors, and producing his magnum opus dealing with matters of authorship and revision of Mary Shelley's *Frankenstein*, especially Percy Shelley's contributions to the text.

Mell observed of Charlie, "[He] did more for the graduate program and the department in general than anyone than any one I can think of during the years we worked together. He was a walking encyclopedia when it came to the graduate students. He knew about their lives, interests, activities, whereabouts, and careers, and cared deeply for their success."[22]

John Ernest, Judge Hugh M. Morris Professor and current chair of the University of Delaware's English department, notes, "Charlie made my life richer, more enjoyable, and deeper from the day I arrived at UD." They would meet at regular intervals for lunch during Ernest's first years at the university, "which always involved a beer-selection process that demanded the same level of Charlie's attention that he gave to everything he cared about, from grocery shopping to authoritative editions of *Frankenstein*." Ernest recalls,

> I was amazed by the records Charlie kept on our graduate program's alumni. He kept track of individual careers, from their publications to their moves from one institution to another, and he continued to offer advice, send information about open positions, and call for updates on their lives. This kind of individualized attention—joined with endless lists, an encyclopedic knowledge of departmental history, detailed attention to any development in the library, and endless email mailings of selected articles from the *Chronicle of Higher Education* and updates on professional opportunities—is exceptionally rare in higher education, but it is just the way Charlie approached everything.

Moreover, according to Ernest, "anyone who was smart enough to let him into his or her life would find a devoted friend and tireless advocate," and "there was no one remotely like him, though we would all benefit from trying to learn from his exemplary approach to life."[23]

Both Ernest and Mell remember Charlie's spontaneous, jovial authorship of limericks, which Charlie would often compose and read at committee and department meetings to relieve tension. To Mell, these "incomparably humorous and witty" ad hoc contributions "reminded his colleagues of the brighter sides of academic life." Ernest recalls a unique tradition that Charlie started at UD for colleagues retiring from the department. Every retiring colleague was "graced with a parting limerick" composed by Charlie and read by him "with considerable drama" at the department's end-of-year reception. When Charlie was nearing retirement, Ernest asked himself who would write Charlie's limerick. By that time, Ernest recalls, "he had become such an important part of my life that I took up the challenge myself—but instead of a poem, I wrote him a song, which I recorded and played at the department's reception." The refrain from Ernest's tribute concludes this memorial postscript:

> Have you met Charlie? I'm sure that he'll know
> He keeps a list of everything as far back as it goes
> Some have tried to trip him up, but I would never dare
> He's the keeper of the records
> He's the bard of Delaware.

NOTES

1. All remarks by contributors in quotations refer to their email messages to the editor.

2. See Jeffery Vail, "In Memoriam: Charles E. Robinson," *Byron Journal* 45, no. 1 (2017): 1–2; and Susan J. Wolfson, "Charles E. Robinson (1941–2016)," *Keats-Shelley Journal* 66 (2017): 25–28.

3. Charles E. Robinson, *Byron and Shelley: The Snake and Eagle Wreathed in Flight* (Baltimore: Johns Hopkins University Press, 1976); Robinson, ed., *Mary Shelley: Collected Tales and Stories* (Baltimore: Johns Hopkins University Press, 1976).

4. Recent noteworthy, ongoing work to challenge this manner of conscious and explicitly biased canonization in the field includes the social media collective and research network Bigger 6 Romanticism. See Deanna Koretsky and Joel Pace, guest eds., "The Bigger Six Collective: Coda: From Coteries to Collectives," *Symbiosis: Transatlantic Literary & Cultural Relations* 23, no. 1 (Spring 2019): 139–140.

5. Charles E. Robinson, ed., *Lord Byron and His Contemporaries: Essays from the Sixth International Byron Seminar* (Newark: University of Delaware Press, 1982); Charles E. Robinson, Bernard Beatty, and Tony Howe, eds., *Liberty and Poetic License: New Essays on Byron* (Liverpool, UK: Liverpool University Press, 1988).

6. *The Mary Shelley Reader*, ed. Charles E. Robinson and Betty T. Bennett (New York: Oxford University Press, 1990).

7. *Mary Shelley's Mythological Dramas: "Proserpine" and "Midas,"* ed. Charles E. Robinson and Betty T. Bennett (New York: Garland, 1992).

8. *The Frankenstein Notebooks: A Facsimile Edition of Mary Shelley's Novel, 1816–17 (with Alterations in the hand of Percy Bysshe Shelley) as it Survives in Draft and Fair Copy Deposited by Lord Abinger in the Bodleian Library, Oxford*, ed. Charles E. Robinson, 2 vols. (New York: Garland, 1996). The *Notebooks* concludes the 9 vol. series *The Manuscripts of the Younger Romantics*, general ed. Donald H. Reiman (New York: Garland, 1985–1996).

9. *Frankenstein: or, The Modern Prometheus: The Original Two-Volume Novel of 1816–1817 from the Bodleian Library Manuscripts. By Mary Wollstonecraft Shelley (with Percy Bysshe Shelley)*, ed. Charles E. Robinson (New York: Oxford University Press, 2008).

10. David Guston, email message to editor, January 28, 2020.

11. Marsha Manns, email message to editor, October 15, 2018.

12. Lisbeth Chapin, email message to editor, August 19, 2018.

13. Jeffery Vail, email message to editor, October 5, 2018.

14. L. Adam Mekler, email message to editor, June 27, 2018.

15. Halina Adams, email message to editor, July 10, 2018.

16. Anna Mercer, email message to editor, November 27, 2018.

17. Geoffrey Bond, email message to editor, July 19, 2018.

18. Andrew Stauffer, email message to editor, July 15, 2018.

19. Marilyn Gaull, email message to editor, July 9, 2018.

20. Hermione deAlmeida, email message to editor, August 27, 2018.

21. Doucet Devin Fischer, email message to editor, August 22, 2018.

22. Donald Mell, email message to editor, August 6, 2018.

23. John Ernest, email message to editor, July 31, 2018.

Notes on Contributors

Brian Bates teaches at California Polytechnic State University, in San Luis Obispo. He completed his undergraduate studies at the University of Delaware under the direction of Charles Robinson. His research focuses on print and performance cultures in eighteenth- and nineteenth-century literature. Bates has published research and criticism on several British Romantic poets. He is the editor of *Keats in Popular Culture* for the *Romantic Circles* Praxis Series and the author of *Wordsworth's Poetic Collections, Supplementary Writing and Parodic Reception*.

Lisbeth Chapin is an associate professor of English at Gwynedd Mercy University, where she teaches courses in British literature, literary theory, literature and the environment, and the Bible as literature. She was a student of Charles Robinson, whose guidance aided the completion of her doctoral dissertation. Chapin's research interests include the materiality of nature in the works of Percy and Mary Shelley. She has published essays on Percy Shelley's poetry and prose, and her most recent publication, "Mary Wollstonecraft Shelley: *Frankenstein*, Grief, and Graves," appears in the Gravestone Project.

Lisa Crafton is a professor of English at the University of West Georgia, where she teaches on Romanticism and British women writers. She is the editor of *The French Revolution Debate in English Literature* and the author of *Transgressive Theatricality, British Romanticism, and Wollstonecraft*. Crafton has also written numerous essays on Wollstonecraft, Blake, and Wordsworth. Her most recent essay, "'Tangle of Matter and Ghost': U2, Leonard Cohen, and Blake in Romanticism," appears in the volume *Rock and Romanticism*.

Robin Hammerman is a teaching associate professor at the Stevens Institute of Technology, where she teaches courses in literature and communications. She is the editor of *Womanhood in Anglophone Literary Culture: Nineteenth and Twentieth Century Perspectives* and, with Andrew Russell, of *Ada's Legacy: Cultures of Computing from the Victorian to the Digital Age.* Her research and teaching interests include the intersections of literature, science, and technology. Hammerman's acquaintance with Charles Robinson spans over two decades and grew significantly during their several years of service to the Byron Society of America, for which she was the director of membership and academic services.

MARK A. MCCUTCHEON is a professor of literary studies at Athabasca University. He is the author of *The Medium Is the Monster: Canadian Adaptations of Frankenstein and the Discourse of Technology.* He has published research and criticism on Romanticism, Canadian popular culture, and copyright policy. McCutcheon is also a writer of poetry and short fiction. His creative work appears in several literary magazines, including *EVENT, Existere, Carousel,* and *subTerrain.*

L. ADAM MEKLER is an associate professor of English and codirector of the freshman English program at Morgan State University. He completed his undergraduate studies at the University of Delaware under the direction of Charles Robinson. In addition to teaching composition and literature courses at both the undergraduate and graduate levels, Mekler has published essays on the works of Mary Shelley, Jane Austen, and Lord Byron in journals including *Studies in Romanticism* and the *Modern Critical Views* series. He is the coeditor of *Mary Shelley: Her Circle and Her Contemporaries.*

SIOBHAN WATTERS is a PhD candidate at Simon Fraser University's School of Communications, where she also teaches. She holds a bachelor's degree in English from the University of Waterloo and a master's degree in theory and criticism from the University of Western Ontario. Her research focuses on food as a technical object and the role it plays within wider technological and economic systems. Her theoretical framework is informed by the philosophy of technology, media archaeology, and Marxist political economy.

SUSAN J. WOLFSON, a professor of English at Princeton University, is widely published in the field of British Romanticism—most recently in *Romantic Studies and Shadows,* which includes pages on *Frankenstein.* She is the editor of the widely admired Longman Cultural Edition of *Frankenstein* and, with Ronald Levao, the editor of the prize-winning *Annotated Frankenstein.*

Index